NEW

THINKING

=

About the Editor

Louise Hodgson is programme director of The Undergraduate Awards (UA), an initiative founded in Ireland in 2008 to celebrate the brightest students who propose fresh arguments and new approaches in their studies. From humble beginnings, The Undergraduate Awards has rapidly expanded into hundreds of universities across the globe, and receives the official patronage of the President of Ireland, Michael D. Higgins. The annual UA Global Summit, which involves workshops and think-ins with these bright young minds, has been described as 'a Davos for students'.

With a background in the media, Louise was previously editor of *Who's Who in Irish Business* 2008–2010; *Most Influential US-Irish Business Leaders 2008*; and *Life Sciences Review*. She lives in Dublin.

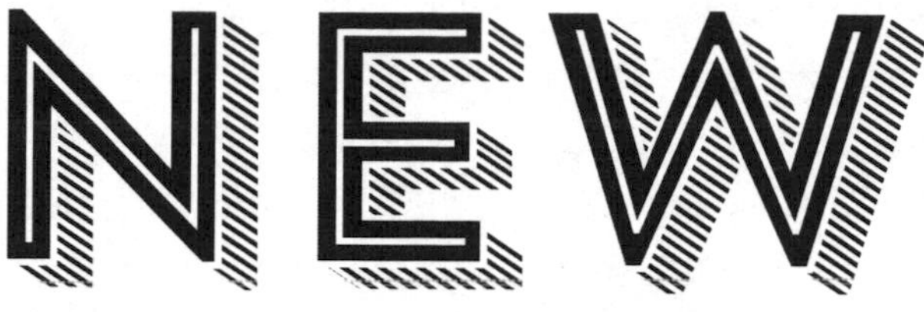

THINKING

=

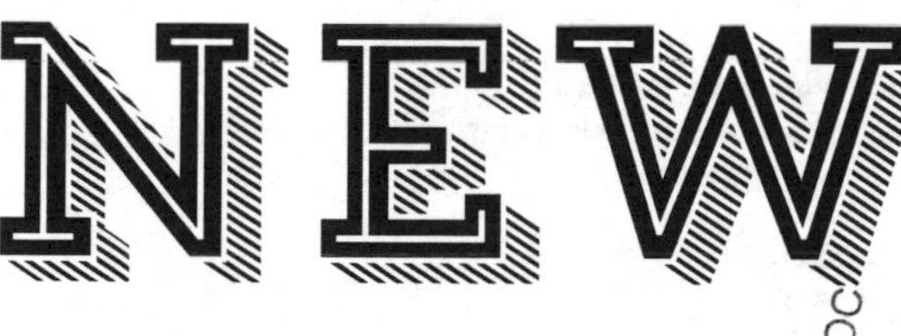

IRELAND

EDITED BY LOUISE HODGSON

Gill & Macmillan

Gill & Macmillan
Hume Avenue, Park West, Dublin 12
with associated companies throughout the world
www.gillmacmillanbooks.ie

978 07171 5854 6

Typography design by Make Communication
Print origination by Síofra Murphy
Printed and bound by GraphyCems, Spain

This book is typeset in Minion and Neue Helvetica.

The paper used in this book comes from the wood pulp of managed forests. For every tree felled, at least one tree is planted, thereby renewing natural resources.

A CIP catalogue record for this book is available from the British Library.

5 4 3 2 1

Contents

Chapter 1

Louise Hodgson

Introduction: Why This Book Is Important

My mother always accuses me of being ageist – and this book probably heralds the end of any argument I had to the contrary. The oldest person I approached to contribute to the book is only 31 (the youngest is 21) and I can't lie: that was on purpose.

For the last three years I have run The Undergraduate Awards (UA), a Dublin-based academic initiative that has grown substantially from first operating in the nine universities on the island of Ireland in the academic year 2008/2009. Today, UA runs two awards programmes: one throughout all third-level institutions on the island and one involving nearly a hundred of the world's most prominent universities. Every year we identify the world's top students and bring them together with Ireland's top students at a summit event in Dublin in November.

One of the reasons I genuinely adore what I do is because it has opened up to me a world of remarkably intelligent, inspiring and (yes) *young* individuals. I have been fortunate to meet, work with and award many of these young people. Take UA's founders, for example. Paddy Cosgrave and Oisin Hanrahan were in their mid-twenties when they started UA five years ago. Since then, more initiatives have cropped up with that same aim of identifying, recognising, encouraging, supporting, celebrating and connecting young people from around the world, usually with an age limit of 30.

I estimate that there are hundreds of these networks that welcome 'bright young things', 'young entrepreneurs' and 'leaders of tomorrow' and offer them a global network of like-minded individuals and a further accolade to add to their CVs. And so there should be! There is nothing more powerful (and yet so simple) than telling people that they are great at what they do – and the sooner this can be done in someone's adult development the better. UA was built on the principle that a reward needn't be much more than a pat on the back and a gold medal; these things are enough to sufficiently encourage a talented individual. (Although, thankfully, UA also offers a lot more besides.)

All of the contributors to this book are associated with some kind of network (if not two or three of them) in operation in Ireland today. The largest of these networks is the Global Shapers Community. It is supported by the World Economic Forum and its Dublin hub has about twenty-five members. These 'shapers' (a title that admittedly works better in other countries) are individuals 'who are exceptional in their potential'. I have been a member of the Dublin hub since it was first launched in 2011 – although I got in by the skin of my teeth three weeks before my thirtieth birthday. (Members must be under 30 on entering, but since I made that deadline, they'll let me hang around until I'm 33.)

In early 2012 John Egan (the current co-curator of the Dublin Global Shapers hub, who appears in Chapter 4) brought the Sandbox Network to Dublin. These 'sandboxers' are young leaders and the network is all about creating lasting connections between them years before they go on to achieve their predetermined success. And then, of course, there are past UA winners, who are some of the smartest graduates Ireland has produced in the last five years. These are the researchers, the thinkers and the doers, who are quickly expanding the UA Alumni Network all around the world.

Members of these networks in Ireland are of the generation that has matured during the recession. A lot of the authors in this book either entered into or graduated from college when the recession first hit, i.e. when the jobs market changed dramatically. It's fair to say that (for understandable reasons) most of the media attention on this generation has focused on their misfortunes. The predominant message to these young people has been that there are no jobs, everyone's leaving the country and they will be burdened with debt for many years to come.

Unfortunately, there is a lot of truth in that message. It would be foolish to say otherwise. However, hearing the message over and over again is depressing. Daily news reports that Ireland's recession has worsened again because of some factor or other (slow EU recovery, a revised economic forecast, etc.) feel a bit like news flashes telling us that a dead horse now has a broken leg. The Irish economy as we know it is broken. We are well aware of this. The newsflashes that tell us the economy is a little more broken than originally anticipated don't serve much purpose. After a while, it starts to sound like white noise.

This is not to say that we shouldn't be appropriately informed of the exact state of affairs: obviously, that is incredibly important. Nor does it mean that the people who played an immoral role in Ireland's downfall should not be held accountable. We cannot brush aside the enormity of the recession; however, it cannot be all we hear about as we aim to recover. As anyone who has ever faced any difficult challenge can attest: a positive mind-set is half the battle. No one ever climbed Mount Everest thinking they were too weak to do it.

Any successful person will delight in telling you that failure is good. Failure makes you learn. When viewed correctly, failure is a wake-up call; it can actually empower you and encourage you. Ireland has failed. We've been knocked down. We need confidence to get back up.

I believe this confidence exists among individuals in many areas. The group I'm closest to is the group of young people who are making a difference. There are twenty of them in this book, but many, many more exist. They are powering ahead when others might give up. They see problems, rename them 'challenges' (much easier to face) and turn them into opportunities. And they do this across all areas of Irish life, society and – of course – the economy.

It's important to give these people a voice, so I got in touch with people I knew (and others I had heard about) and asked them this: 'What is your vision for Ireland?' I told them it could be anything, from mad, wacky ideas, to real, evidence-based proposals. This was a brainstorming session: there were no wrong answers.

This book features an interesting mix of people. Big-thinking entrepreneurs feature alongside PhD candidates who've spent years focusing on refined areas of research.

What really matters is that each one is as passionate as the other about seeing Ireland recover. Julie Clarke examines cost-efficient tunnelling, Roslyn Steer imagines a 'music utopia' and Tara Duggan explains effective composting. Matthew Smyth aims to turn all Irish people into avid theatre-goers, Bella FitzPatrick proposes that everyone should learn how to knit and Eimhin Walsh states the case for rationality in religion. Some contributors depart from their area of expertise and write instead about something they genuinely want to address. Sasha de Marigny encourages us to celebrate all nationalities that come to live in Ireland. Éabha Ní Laoghaire Nic Ghiolla Phádraig (writing through the Irish language) simply wants to remind us that we've faced challenges in the past and we have overcome them, as evidenced in the literature of the great Irish authors and poets.

UA is all about inter-disciplinary collaboration: it is unique in being the only academic awards programme operating internationally that recognises students across all disciplines. And yet, in all the different essays featured in this book, similar themes emerge. The most prominent one is the issue of emigration. That's no big surprise – the departure of tens of thousands of people, mostly young, over the last five years has been one of the most striking effects of the recession. What will be surprising to some readers, though, is the attitude the contributors of this book have towards this recent development. In these pages at least, emigration isn't necessarily a bad thing. Many of the contributors have experience of emigration. Orla Power studied in London and now lives in New York, as does Oisin Hanrahan. Dublin-based Sasha de Marigny is from South Africa – her mother originally moved there from Galway. William Peat runs NGen with an Irish co-founder who is based in San Francisco. It is clear from these contributors that there are many opportunities brought about by such international movement of people: opportunities for the individual and for both countries involved.

It could be expected that the theme of emigration would appear in William Peat's essay on Diaspora. However, it features in the essays of many other contributors: Padraig Mannion (Careers), Matthew Smyth (Theatre), Darren Ryan (Social Entrepreneurship) and Orla Power (Banking II). In each case, it is viewed in a positive light. The world is not the place it was during the Great Famine, nor is it the place it was during the 1980s. Thanks to Facebook, Twitter, Skype and WhatsApp (not to mention many thousands of international airline operators) you can move to the other side of the world and still stay connected. Indeed, 'home' has become rather an enigmatic concept and experiencing life on the other side of the world (or simply across the Irish Sea) can be so exciting. When Ireland recovers, people will still emigrate and those who have already emigrated may return – or they may not, in which case they'll join the incredible diaspora that has benefited our country in so many ways.

In addition to viewing emigration in positive terms, some of our contributors believe that the recession itself is not entirely a bad thing. In creative circles, a recession can mean a spurt of activity. Katie Tsouros (Art), Matthew Smyth (Theatre) and Roslyn Steer (Music) stipulate that a downturn (i.e. a lack of resources) is no obstacle to the arts. As Roslyn Steer says, 'If the next generation of artists is not burning with frustration, there's probably something wrong.'

Humour aside, it has to be noted that artists are good at finding creative solutions to the challenges they face and this is why they tend to flourish in times of hardship. That's not to say that hardship is good; however, because of the cyclical nature of economics, hardship is inevitable. It is certainly very real in Ireland at the moment and we could do with some creative solutions now.

Social media and technology are among the other prominent themes in this book: Google, Facebook and LinkedIn are referenced repeatedly. It is interesting to consider the ways in which young Irish people have been impacted by the fact that Ireland hosts major operations of these global companies. The essays in this book would lead us to believe that this is an aspect of Irish society of which most young people are proud. These global giants are shaping and changing the way our world works. They are leading the online revolution that extends to the smallest of tech start-ups hacking together apps from bunk beds in Silicon Valley. And they are employing thousands of people in Ireland today. (Interestingly, it was UA's co-founder Paddy Cosgrave who set up the Web Summit series. The Dublin event of the series brings hundreds of technology professionals to Ireland and it is one of Europe's most celebrated technology conferences.)

Perhaps inevitably, banking and economics feature in several of the essays. There are two essays on banking and each of the contributors takes their own approach on the subject. Orla Power, a young woman working in banking in one of the most critical cities for global financial processing, writes from the heart. She talks about the people – the bankers – who are now tasked with rebuilding trust in a system that failed so many. John Egan, on the other hand, looks beyond recovery and paints a thrilling picture of a new kind of banking developed and shaped by the advent of online technology and social networking. There are also two essays on the subject of economics. Robert Nielsen proposes a plan to end the recession and Oisin Hanrahan develops the concept of the Startup Economy, calling on the government to adopt the proven processes of successful young companies to ensure economic growth.

It is evident from this book that *this generation wants to help*. For all the austerity, inadequate government action, lack of funding, depressing headlines and lost people, we have to remember that Ireland has a lot to be proud of. We have produced world-renowned scientists, artists, human rights activists, businesspeople and philosophers. We split the atom, discovered pulsars, founded modern chemistry and explained why the sky is blue. We wrote *Ulysses*, *Gulliver's Travels*, *Dracula*, *The Importance of Being Earnest* and 'With or Without You'. We produced Boyle's law, the Beaufort Scale, the Kelvin Scale and the Dublin Method. We grow the best grass in the world! Katie Taylor, Daniel Day-Lewis, Saoirse Ronan, John O'Farrell, Lorraine Twohill, Brian O'Driscoll, Anne Enright, Seamus Heaney, Mary Robinson, T.K. Whitaker – all wonderful role models for the next generation of world-class athletes, entertainers, entrepreneurs and activists.

Taking stock of what we have already done will enable us to think about what we can achieve in the future. This book is about looking up, around and then forward. The contributors have proposed their 'bright ideas', their blueprint for tomorrow. Not every idea will be practical; not every idea should be implemented. The purpose of proposing these ideas is to get us to think about how Ireland could recover and be a world leader. As a nation, we need to think about how we can continue to influence in areas such as science, technology, peacekeeping and the creative arts. Young people can bring about important change – and they want their voices to be heard.

For a short time at least, forget the dead horse. Let's imagine a new horse for ourselves – a fast, strong, majestic one. Let's try to see the Ireland we can become. Let's look to the future, because the future is bright.

Chapter 2

Éabha Ní Laoghaire Nic Ghiolla Phádraig

Fís na hÉireann

Nuair a bhí Éabha ina páiste, do bhí fonn uirthi a bheith ina naomh proifisiúnta. Nuair a thuig sí nach mbeadh sí in ann a leithéid a bhaint amach, rinne sí cinneadh chun a bheith mar an chéad bhean Éireannach sa Spás amuigh. Dála an scéil, is ea a grá na teangan a fuair an lámh in uachtar ar deireadh agus mar sin ghabh sí leis an bhFraincis agus leis an nGaeilge in Ollscoil Luimnigh, áit ar bhain sí céim chéad ónóracha amach sa bhliain 2012. Anois tá sí ag gabháil le Máistreacht sa bhFiontraíocht i gColáiste na hOllscoile Corcaigh. Deireann sí gur 'thit sí i ngrá leis an nGaeilge trí thimpiste' ar chúis a huncail agus feitheoir den scoth a bhí aici ar choláiste. Anuas air sin, fuair sí ardmholadh mar iarrthóir sna Dámhachtainí Fochéime 2012 sa chatagóir na Gaeilge.

Fuair mo sheanmháthair Hancy Fitzpatrick bás cúpla lá ó shin, beannacht Dé dílis lena hanam agus le hanamacha na marbh, cúpla seachtain sular shroich sí nócha seacht mbliana d'aois. Bhí saol fíochmhar deacair ag mo sheanmháthair, scriosadh gnó a muintire sna 30í agus an Cogadh Eacnamaíoch ar siúl, fuair a fear céile bás agus fágadh í le naonúr páiste idir naíonáin agus dhéagóirí. D'oibrigh sí san ospidéal i rith na hoíche agus thagadh sí abhaile ar maidin chun an bhricfeasta a ullmhú dos na páistí, bhíodh ceithre uair a chloig codlata aici, bhíodh sí éirithe aríst chun an dinnéir a ullmhú agus bhí uirthi dul amach ag obair aríst ansan. Ar ámharaí an tsaoil, áfach, do bhí tacaíocht na gcomharsan aici.

In ainneoin a cuid deacrachtaí, bhí doras an tí i gcónaí ar oscailt agus do bhí fáilte roimh gach éinne ann. Bhíodh cóisirí ar siúl go minic i dtigh mo sheanmháthair. Ní hamháin é sin ach do bhí mo sheanmháthair ina ball bunaidh de chumann na mbaintreach i gCorcaigh le Maureen Black agus ina ball bunaidh d'ionad pobail na Linne Duibhe – seirbhísí sláinte a sholáthar don phobal a bhí mar aidhm acu.

Anois le linn cúlú eacnamaíochta, cuirimse an cheist seo orm féin, 'Cad a dhéanfadh mo sheanmháthair agus na sinsir a bhí ann romham chun déileála le fadhbanna na tíre agus chun réitigh a fháil?' Anuas air seo, braithim go bhfuil freagra na ceiste sin le fáil i litríocht na Gaeilge. Is í an cheist atá ann ná: an bhfuil saíocht agus eagna le fáil i litríocht na Gaeilge a thabharfadh treoir agus spreagadh dúinn cursaí na tíre a chur i gceart aríst? San aiste seo déanfaidh mé tagairt do roinnt saothar Gaeilge a léiríonn ceachtanna tábhachtacha dar liom féin a bheadh fóinteach dúinn maidir le fís dhearfach dheimhneach den tír a chur chun cinn.

CUMHACHT LITRÍOCHT NA GAEILGE

Is cuimhin liom go minic i lár léachtaí litríochta na Gaeilge, bhí orm na deora a shlogadh siar mar gheall ar na mothúcháin cumhachtacha a bhí á nochtadh; ar nós pian, sceimhle, crógacht, neart nó dánaíocht an duine a chuireadh in iúl sa litríocht. Ní hamháin é sin ach go minic bhí fonn gáire chomh láidir orm go raibh na deora ag sileadh amach as mo shúile. Téann litríocht na Gaeilge i bhfeidhm go mór orm i gcónaí, in ainneoin an mheasa atá agam ar litríocht na Fraince agus litríocht an Bhéarla i leith mo chuid léinn, tá gaol is meas amhra agam ar litríocht na Gaeilge mar is léir go mbaineann litríocht na Gaeilge lenár gcuid staire agus lenár sinsir.

Gan amhras tagann féiniúlacht mhuintir na hÉireann mar phríomhthéama chun cinn i litríocht na Gaeilge agus dá bharr san táim in ann stair agus eispéireis mo shinsir agus meon na nÉireannach a thuiscint níos fearr. Ceann de na tréithe a thaitníonn go mór liom faoi litríocht na Gaeilge ná go ndíríonn roinnt mhaith saothar ar an ngnáthduine, agus ceann de na nithe is suimiúla fútha ná go raibh dearcadh agus meon neamhchoitianta, amhra agus speisialta acu. Mar sin creidim féin go bhfuil an-chuid ceachtanna le foghlaim astu.

TÁBHACHT AN PHOBAIL

Léiríonn saol an Bhlascaoid an tábhacht a bhain le cúrsaí an phobail agus na seiftiúlachta ag deireadh an naoú haois déag agus tosach an fichiú haois sa dírbheathaisnéis *An t-Oileánach* le Tomás Ó Criomhthain. Is léir ón bhfrása 'Níl ann ach an fhírinne; níor ghá dom aon cheapadóireacht'[1] go dtugann Ó Criomhthain léargas barántúil agus an-chruinn dúinn ar shaol an Bhlascaoid. Caithfear a thuiscint chomh maith, i gcónaí i dtéarmaí an phobail a labhraíonn Tomás, mar a léiríonn an tagairt seo dúinn: 'Thugas iarracht ar mheon na ndaoine do bhí im thimpeall do chur síos chuin go mbeadh a dtuairisc'.[2] Dá bharr san, bíonn Ó Criomthain ag caitheamh solais ar chruatan an tsaoil ar an oileán: 'Oíche mhór fhada fhuar mar seo ag comharc na mara, go lánmhinic ar bheagán fáltais'.[3]

Do thuig muintir an oileáin go maith cé chomh tábhachtach is a bhí cursaí pobail agus cursaí seiftiúlachta chun maireachtaint beo agus chun feidhmiú mar phobal rathúil. Déanann Ó Criomhthain cur síos ar eachtra áirithe, 'An Long Gail is an t-Arm', inar tháinig saighdiúirí go dtí an oileán ag iarraidh cíos a bhailiú uathu. Tá eagla an domhain ar mhuintir an oileáin go goideadh a gcuid ainmhithe agus go scriosadh a gcuid tithe. Láithreach ball tagann an pobal le chéile chun cosc a chur ar na saighdiúirí agus chun saol an Bhlaoscaoid a chaomhnú: '… fuaireas post ós na mná do bhí bailithe ann. Ag bailiú chloch do cuireadh me agus gach nduine eile do bhí ann'.[4] D'imigh na fir as an áit, bailíodh na mná is na páistí clocha agus chuaigh na mná is na páistí chuig an trá: 'Ná raibh na mná ag fágáil na háite'.[5] Gan dabht ar bith, léiríonn an eachtra seo dúinn neart agus seiftiúlacht na mban! Ní raibh na mná toilteanach a oileán muirneach a fhágáint gan iarracht a dhéanamh é a chosaint.

Ní hamháin go raibh trí báid lán do shaighdiúirí armtha ag teacht go dtí an trá, bheadh na saighdúirí seo in ann robach a dhéanamh den oileán. In ainneoin an bhaoil a bhí ann, 'ní raibh eagla ar na mnáibh rompu'.[6] Dála an scéil, thosaigh na mná agus na páistí ag caitheamh cloiche leo, agus gortaíodh na saighdiúirí go dona. Ar chúis mire bhí bean amháin chomh daingean diongbháilte cosc a chur leis na saighdiúirí, beagnach gur chaith sí a leanbh féin ar son an oileáin: '… go gcaithead an leanbh leo!'.[7] Gan mórán achmhainní agus gan urchair, feidhmníonn muintir an Bhlascaoid mar ghrúpa, mar mheitheal agus mar phobal aontaithe chun maireachtaint beo agus chun saol an Bhlascaoid a chosain, agus mar thoradh air sin: 'Do ghluaisíodar leo [na Saighdiúirí] abhaile insan imeacht go dtánadar, gan bó, gan capall, gan caora'.[8]

Comhoibríonn muintir an oileáin le chéile mar phobal aontaithe agus baintear úsáid as gach uile acmhainn a bhfuil mór-thimpeall orthu chun coisc a chur ar na saighdiúirí. Fiú amháin sa lá atá inniú ann an bhfuil ceacht saoithiúil le foghlaim ón eachtra seo? Tugann sé le fios domsa go bhfuil cumas, neart agus diongbháilteacht ionainn mar phobal agus is léir mar mhuintir aontaithe go bhfuil an fhéidearthacht ionainn cosc a chur ar aon choimhlinteoir a gcuirfeadh isteach ar síth na tíre.

Creidim go bhfuil sé rí-thábhachtach tacaíocht a thabhairt dá chéile agus feidhmiú mar phobal aontaithe. I gcás mo sheanmháthair is é an pobal a thug cabhair agus tacaíocht di. Nochtar dúinn aríst is aríst eile sa stair is sa litríocht gur dream láidir, greamúsach sinn, agus da mba rud é go raibh ár sinsir in ann a leithéid a dhéanamh, gan amhras mar sliochtaigh don dream iontach siúd is féidir linn, agus aontú mar phobal chun fadhbanna agus deacrachtaí eacnamaíochta na tíre a chosc is a réiteach.

IOMPAR SUAITEACH: DÁNAÍOCHT AGUS CLAONAÍOCHT AN DUINE!

Thugadh mo sheanmháthair le fios dom go bhfuil sé riachtanach seasamh ar son cearta an duine, anuas air sin tá sé rí-thábhachtach do ghuth a chur in iúl, i gcoinne noirm agus luachanna na sochaí a chuireann cosc orainn dá bharr, faoi mar a léiríonn Nuala Ní Dhomhnaill ina cuid filíochta.

Sa dán 'Táimid Damanta, a Dheirféaracha', déanann Ní Dhomhnaill tagairt d'Éabha agus tá tagairtí le fáil di i ndánta eile ar nós 'Manach' (1998) agus 'Cnámh' (1988). Braitear sa dán seo gur siombail í Éabha do dhánaíocht agus do chlaonaíocht – is í Éabha foinse pheaca an tsinsir mar rinne sí Ádhamh a mhealladh chun úll Ghairdín Pharthais a ithe. Ach níl an tuairim sin á cur in iúl ag Ní Dhomhnaill in aon chor ach a mhalairt atá idir lámha aici, mothaítear go bhfuil sí ar son na mban claonaí sa dán seo agus ar son shiombail an oilc atá nascaithe le Éabha mar bhean dána agus chlaonaíoch:

Chaitheamar oícheanta ar bhántaibh Párthais
ag ithe úll is róiseanna
laistiar dár gcluasa ag rá amhrán
timpeall tinte cnámh na ngadaithe
ag ól agus ag rangás le mairnéalaigh agus robálaithe.[9]

Ba chuma leis na mná seo sa dán 'Táimid Damanta, a Dheirféaracha' agus tá an bhraistint sin an-láidir agus an-chumhachtach sa dán seo; tugann an t-iompar claonaíoch seo cumhacht do na pearsana sa dán. Is féidir an chumhacht seo a nascadh le dánta eile le Ní Dhomhnaill a chuireann cumhacht na mban agus claonaíocht an duine in iúl dúinn, mar shampla sna dánta 'Labhrann Medb' agus 'Agallamh Na Mór-Ríona Le Cú Chulainn'.

Tá Ní Dhomhnaill ag caitheamh solais ar ghuth an duine agus feictear go bhfuil carachtair na ndánta breá ábalta iad féin a chur in iúl agus tá guth láidir acu, cinnte ní féidir iad a choimeád ina dtost. Faoi mar a deireann Ní Dhomhnaill í féin, '*It is awful to be invisible*'.

Téann na carachtair siúd i bhfeidhm go mór orm agus nochtar dom mar sin go bhfuil gá le iompar suaiteach sa tsochaí agus is cóir dúinn a bheith dána faoi mar atá léirithe ag Ní Dhomhnaill. I lár ghéirchéim éacnamaíochta ina bhfuil tóin na sochaí ag titim as a chéile, an fiú a bheith dána agus claonaíoch? Nach bhfuil gá orainn mar sin labhairt amach i gcoinne na héagóra sóisialta a chuireann cosc lena bhforbairt agus bhforás mar thír dearscnaitheach?

AG CEISTIÚ AGUS AG ATHRÚ NOIRMEACHA SÓISIALTA: SÁRÚ LUACHANNA TRAIDISIÚNTA

Cuireann Ní Dhomhnaill tábhacht dhánaíocht an duine agus iompar suaiteach ós ár gcomhair, agus anuas air sin, léiríonn sí dúinn an géarghá a bhaineann le guth an duine a chur in iúl. Mar thoradh air sin tagann buncheisteanna móra chun cinn agus ceistítear staid na sochaí dá bharr. Sa dán 'Táimid Damanta, a Dheirfeáracha', déanann Ní Dhomhnaill gaol agus nasc idir an carachtar mímhorálta Izeibil agus na mná nach gcloíonn leis an ról agus an stádas a ceadaíodh dóibh sa chóras patrarcach.

Dá dtógfar fealsúnacht an eiseachais san áireamh i leith anailís an dáin seo, agus mar a dúirt Earnshaw, '*Existence precedes essence... man being essentially "nothing" but what he makes of himself*',[10] is féidir linn an nasc a dhéanamh idir fealsúnacht an eiseachais agus an dán seo. Braithim mar sin go bhfuil rian fealsúnacht an eiseachais le fáil sa dán seo.

Go háirithe nochtar dúinn go bhfuil na coincheapa *être-pour-soi* (being-for-itself) agus *transcendance* an-thábhachtach i leith anailís na gcarachtar sa dán seo. Chun sampla a chur in iúl: an duine a chleachtann an coincheap *être-pour-soi*, ní chloítear le rialacha na sochaí má chuireann na rialacha sin cosc ar shaoirse an duine agus dá bharr ní ghlactar le noirm na sochaí ach ceistítear staid na sochaí agus ról nó cumhacht na n-údarás. Dar le Jean-Paul Sartre, tá an dualgas ar an duine freagracht a ghlacadh mar tá freagracht riocht an tsaoil agus staid na sochaí orainn, agus bíonn fios ag an duine go gcaitear glacadh le freagracht as riocht an tsaoil agus staid na sochaí mar thoradh go bhfuil comhfhios an eisidh aige. An aidhm atá taobh thiar den choincheap *être-pour-soi* is ea saoirse an duine a bhaint amach, agus dá bharr caitear freagracht iomlán a ghlacadh as staid agus riocht an tsaoil faoi mar atá sé. Tugann coincheap an eiseachais le fios dúinn nach bhfuil sé sásúil locht a chur ar an sochaí, ar an stát, ar an eaglais ná ar aon institiúid údarásach eile. Léirítear dúinn mar sin i dtéarmaí an choincheapa go bhfuil an chumhacht ionainn agus tá an chumhacht againn chun cursaí an tsaoil a athrú.

Ciallaíonn *transcendance* i dtéarmaí fealsúnacht an eiseachais nach bhfuil an duine teoranta de bharr na haicme sóisialta, róil sóisialta, noirm na sochaí nó luachanna sóisialta. Dar le Sartre is cuma cén stádas nó cén ról atá ag an duine sa tsochaí, fós tá an duine sin in ann an stádas nó an ról atá acu a sharú mar níltear teoranta ar chor ar bith.

Ag déanamh tagartha do coincheapa *être-pour-soi* agus *transcendence* sa dán 'Táimid Damanta, a Dheirféaracha', tá sé le tuiscint againn mar sin nach bhfuil na mná seo ag cloí le struchtúr na sochaí patrarcaí, is cuma leo cén ról atá curtha i bhfeidhm ag an sochaí seo i dtaobh na mban ach níl na mná sa dán seo chun glacadh leis; níltear sásta glacadh le staid na mban nó riocht an tsaoil le haghaidh na mban, 'Ná bheith fanta/Istigh ag baile ag déanamh tae láidir d'fhearaibh'.[11] Ciallaíonn sé sin i dtéarmaí fealsúnacht an eiseachais nach bhfuil na mná seo teoranta de bharr na haicme sóisialta, róil sóisialta, noirm na sochaí nó luachanna sóisialta. Is cuma cén stádas nó cén ról atá ag na mná seo sa sochaí, fós táid in ann an stádas nó an ról atá acu a shárú mar nílid teoranta ar chor ar bith. 'Is rince aonaoir a dhéanamh ar an ngaineamh fliuch',[12] níl siad sásta pléisiúr an tsaoil a chosc dá bharr. Braithtear go bhfuil aoir shóisialta idir lámha ag Ní Dhomhnaill anseo.

Mar tugtar le fios dúinn, cuireann na mná seo a gcuid guthanna bríomhara, athléimneacha agus greannacha in iúl go láidir dúinn. Is mná dochloíte, dáigh agus misniúil iad. Tá sé mar sprioc agus mar aidhm acu saoirse, lúcháir agus eacstais a bhaint amach ina saol, agus dá bharr iarrtar orainn staid na sochaí a cheistiú agus solas a caitheamh ar na noirm agus na luachanna nach oireann dúinn a thuilleadh. Téann teagasc an dáin i bhfeidhm go mór orm mar ar nós na gcarachtar sa dán seo a rinneadh cinneadh comhfhiosach chun athraithe tábhachta a dhéanamh sa tsochaí, braithim go bhfuil an cumas, an acmhainneacht agus an poitéinseal ionainn staid na sochaí agus buncheisteanna móra a chur in iúl, agus mar thoradh air sin is féidir linn cosc a dhéanamh ar an dochar a rinneadh. Is léir dúinn mar sin go mbeadh an deis againn teacht ar fhuascailtí agus ar réitigh agus go mbeadh sochaí chomhchuí ann dá bharr.

BAINIGÍ TAITNEAMH AS: TÁBHACHT AN MHEOIN

Nochtar dúinn i dtosach an úrscéal dírbheathaisnéise *Fiche Blian ag Fás* (1933) le Muiris Ó Súilleabháin in ainneoin chruatan a óige is léir go raibh meon speisialta ag Ó Súilleabháin: 'Ná rabhas ach leathbhliain d'aois nuair a fuair mo mháthair bás ... ní raibh éinne chun aire a thabhairt domsa'.[13] Cuireann sé in iúl dúinn gur bhain sé an-taitneamh as an ngnáthshaol agus as gnáthimeachtaí an lae: 'Ní fada go rabhas i dtaibhreamh, go rabhas féin agus Micil Dé ag siúl trí pháirc leathan i mBaile an Mhuilinn ag baint bláthanna deasa'.[14] Is féidir a rá mar sin go bhfuil dearcadh suaithinseach ag Ó Súilleabháin i leith an tsaoil a bhí mórthimpeall air, dá bharr san tagann fíorshonas an tsaoil amach ina chuid scríbhneoireachta. Anuas air sin, braitear i leith a stíl scríbhneoireachta go bhfuil nasc ann idir mheon dearfach Uí Súilleabháin ina léiríonn sé aoibhneas an tsaoil, agus cruthaitheacht agus nuálacht an duine mar shampla nuair a deireann sé, '"Ó," deireadh sí, "tabharfaidh mé isteach in áit dheas inniu thú." – "Agus an bhfuil milseáin ann?" arsa mise'.[15] Is léir go mbaineann Ó Súilleabháin gach uile deis is caoi amach, fiú amháin ag aois an-óg tugtar le fios dúinn go raibh an-chuid cruthaitheachta agus grinn ag baint leis an tslí ina chaith sé le cúrsaí an tsaoil.

Déanann sé cur síos ar eachtraí éagsúla laethúla a léiríonn cruthaitheacht agus nuálacht an duine agus nochtar iad le dearcadh agus le meon dearfach chomh maith. Tá sé soiléir dúinn gur bhain Ó Suilleabháin agus muintir an Bhlascaoid an-taitneamh as an ngnáthshaol. Sa chaibidil 'Rásieanna Fionntrá' nochtar a leithéid dúinn. Sa chaibidil seo téann Muiris agus Tomás go dtí na rasaí agus cuirtear síos ar eachtra an-ghreannmhar ina d'éiríodh an bheirt acu ar meisce: 'Cá mbeimís ach ar aghaidh thí óil amach i gCeann Trá'.[16] Tosaíodh ag aiseag is ag cuir amach dá bharr agus, saoithiúil go leor, léirítear dúinn soineantacht agus saontacht na mbuachaillí i stíl an-ghreannmhar nuair a bhfuarthas milseáin ar an tsíl abhaile. I rith na caibidile 'Oíche Shamhna', tugann Ó Suilleabháin mionchuntas dúinn ar chuid de na heachtraí spraoiúla a bhí á chruthú acu, ar nós na gcluichí, an fhéilteachais agus na siamsaíochta a bhí ar siúl acu le linn Oíche Shamhna agus 'árdoíche le bheith againn'.[17] Ag déanamh tagartha de na cásanna a luadh, braitear go bhfuil nasc ann idir cruthaitheacht is nuálacht an duine agus pleidhcíocht is greann an duine. Cinnte tógadh na deiseanna a tugadh dóibh chun rud éigint nua a thriailadh agus mar thoradh baineadh sárthaitneamh as! I mo thuairmse is ceacht an-luachmhar atá á nochtadh anseo.

Tugtar le fios dúinn maidir leis na heachtraí a luadh nach raibh cursaí an tsaoil ró-thromchúiseach nó dáiríre in ainneoin chruatan an tsaoil, faoi mar a léiríonn Ó Súilleabháin dúinn. Baineadh taitneamh as an saol! Is dream deisbhéalach, tráthúil agus abartha sinn, cinnte is cuid lárnach é sin do mheon na hÉireann. Anuas air sin, caitheann Ó Súilleabháin solas ar thábhacht meoin agus dearcaidh dearfach i leith aoibhneas an tsaoil; ar nós mo sheanmháthar agus na cóisirí a bhíodh ar siúl ina tighín beag. In ainneoin chruatan an tsaoil, is fiú dearcadh suaithinseach a choiméad. Bainigí lán-taitnimh as an saol!

Ní amháin go bhfuil na ceachtanna a luadh le fáil i litríocht na Gaeilge ach b'iad siúd na ceachtanna a bhíodh ionchollaithe ag mo sheanmháthair go dtí lá a báis agus í ina sin-sin-seanmháthair. Sin í an fhís a bhíodh aici don tír seo: go mbeadh na pobail ag comhtháthú le chéile, go gcuirfí béim ar thábhacht na seiftiúlachta agus go n-úsáidfí na hacmhainní atá ann mórthimpeall orainn, gan gearán a dhéanamh faoi na heaspa acmhainní nach bhfuil ar fáil ar nós mhuintir an Bhlascaoid.

Dála filíochta Nuala Ní Dhomhnaill bígí dána agus claonaíoch. 'Ná glac leis an saol ná le staid na tíre muna n-oireann sé duit agus muna bhfuil tú sásta leis', chloisfinn mo sheanmháthair á rá liom. 'Ní fiú gearán amháin faoi muna ndéanfadh sé aon mhaitheas d'éinne eile', a déarfadh sí liom, 'caithfidh tú do ghuth féin a chur in iúl – labhair amach!' Ar nós na mban sa dán 'Táimid Damanta, a Dheirféaracha', má tá daoine ag fulaingt tá gá noirm sóisilata a ceistiú agus má tá gá mar sin caithfear iad a athrú.

D'aontódh sí le meon aigne Uí Súilleabháin: nach bhfuil cursaí an tsaoil ró-thromchúiseach nó ró-dháiríre, 'mar sin bainigí taitneamh as an saol, agus tiocfaidh cruthaitheacht agus nuálacht an duine chun cinn dá bharr', a déarfadh sí liom.

Anois braithim macalla focal mo sheanmháthair ag athshondach ionam. Do thuig sí siúd go deimhin an cumas, an acmhainneacht agus an poitéinseal atá ionainn chun fuascailtí agus réitigh a fheabhsóidh staid na tíre a bhaint amach agus a cruthóidh sochaí níos comhchuí dúinn agus dóibh siúd atá fós le teacht.

Notes

1 Ó Criomhthain, T. (2002), *An t-Oileánach*. Baile Átha Cliath: Cló Talbóid, p.325.
2 Ibid. 327.
3 Ibid.
4 Ibid. 50.
5 Ibid. 51.
6 Ibid.
7 Ibid. 52.
8 Ibid. 53.
9 Ní Dhomhnaill, N. (1988), *Selected Poems: Rogha Dánta*, tr. Michael Hartnett. Baile Átha Cliath: New Island, p.14.
10 Earnshaw, S. (2006), *Existentialism: A Guide for the Perplexed*. London: Continuum International Publishing Group, p.74.
11 Ní Dhomhnaill, N. (1988), *Selected Poems: Rogha Dánta*, tr. Michael Hartnett. Baile Átha Cliath: New Island, p.15.
12 Ibid.
13 Ó Súilleabháin, M. (2011), *Fiche Bliain ag Fás*. An Daingean: An Sagart, p.11.
14 Ibid. 19.

15 Ibid. 11.
16 Ibid. 74.
17 Ibid. 54.

Chapter 3

Katie Tsouros

Art

Katie is an art curator specialising in the emerging art market. In 2010, having just achieved her MA in Contemporary Art from Sotheby's Institute of Art, London, she founded KTcontemporary, a contemporary art gallery based in Dublin and dedicated to exhibiting work by emerging graduate artists. Flexing her entrepreneurial muscles even further, Katie recently launched Artfetch, a curated ecommerce platform. Artfetch selects the most promising emerging artists from around the world and presents their work for sale online. As co-ambassador of the Dublin Sandbox hub and a WEF global shaper, Katie was the obvious choice to lead the judging panel of the first ever Visual Arts category of The Undergraduate Awards in 2012.

Let's begin by asking, why does art matter? In a country that is struggling to survive economically, why should we worry about the future of art? Why should we care?

Since the beginning of time art has captured the contemporary zeitgeist. It encapsulates a particular moment, a sentiment and feeling that visually represents and reflects the current state of society. Before the camera was invented, this was quite literal. Art, be it ancient cave drawings or Renaissance painting, was used to chronicle events, record happenings, people and experiences, portray places, share memories and observe and remark on the status of culture, civilisation and humanity. Of course there has always been an element of pure decorative countenance, but even that echoes a timely response. In this respect, little has changed: the medium of expression has broadened and the remit for communicative freedom has augmented, but the basis for creation remains largely the same. Take away art and you take away a neutral discourse and an innate narrative of our time.

Art has the power to change the way we view the world, the power to make us stop, consider and think about an alternative perspective. It has the power to challenge, investigate and explore another position; to question our perceptions and realities. My journey through art has led me to my interests in visual culture and entrepreneurship – two things I believe can be very closely linked. When I finished school I wanted to study business and economics at university. At the time, I loved art – not *making* it (I was terrible at that) but learning from it, reading about it, discovering the circumstances that surrounded the realisation of it. And, of course, I loved seeing it; seeing these great pieces in real life that beautifully depicted a wealth of knowledge and a certain vantage point, aesthetically satisfying, encompassing and heart-hitting.

Simply put: art enriches our lives.

Just after my Leaving Certificate I went to visit the Venice Biennale, the most prestigious global contemporary art exhibition, and arguably the most important. When I was there I realised this was something that I wanted to be a part of in a very real way. I wanted to be part of something that I felt was impactful and meaningful. Something in me knew that this, the medium of art, was how I was going to learn the most that I could about the world. When I returned home I changed my university applications and set out on a new path that saw me through a BA in Art History and Philosophy at University College Dublin and an MA in Contemporary Art from Sotheby's Institute of Art, London. Throughout and following that time I worked at various galleries and institutions both in Dublin and London and my path eventually (maybe inevitably, though I had no idea at the time) led me back to setting up my own art businesses.

The late great genius Andy Warhol said, 'Good business is the best art.' My experience straddling both the business and art worlds has taught me more than I could imagine, influencing my outlook on both sides of the table. In fact, the similarities between artistic endeavour and entrepreneurship are, perhaps somewhat surprisingly, intrinsically connected (though professionals in both arenas might be reluctant to admit that). A lot can be learned from methodologies applied to artistic practice; the characteristics and traits of artists and entrepreneurs can be paralleled. Most noticeable is their inherent creativity and innovative approach; their capacity for making, connecting, observing, integrating and appropriating; their aptitude for flexibility and improvisation; and their talent to conceptualise an unfolding tale. Both artists and entrepreneurs see the gaps in the world and try to fill them.

Herein lies the future of art in Ireland. Now that we've determined the value of art and artists, what can we learn from their innovative approach and how can we nurture that in this country? It begins by *perpetuating* the future of art and artists, cultivating a culture of appreciation and establishing a sense of self and identity for Irish contemporary art that can lead to real longevity. Comparably, how can we apply methodologies of business to achieve that?

It is difficult to talk about any strategy for permanence in the arts without some national or government input. The lack of a collecting institution is a great disadvantage. Without a public museum to actively collect, document, archive and educate (this piece is key) for the next generation of artists, it's hard to establish any sort of weight or international recognition – support and endorsement should always begin at home. Ireland is a young state and the Irish Museum of Modern Art – the primary body for the assemblage and presentation of modern and contemporary art – was founded as late as 1990. Less than twenty-five years later, its function has all but dried up. With an absent spending budget to buy and house the work of emerging and established artists, its purpose is somewhat lacking. After all, if an artist is not collected by their national institution, how can we expect them to be collected elsewhere? Granted, the US model for institutional patronage is devoid (more or less) of any government input, relying solely on corporate donations and individuals' discretion – but this in itself opens up a whole host of problems, and resources of this nature are limited on our small island.

In Europe, general privatisation within the arts has never been the way to go, making public engagement much more accessible and successful. Germany is a prime example of how cultivating a government-led initiative for the collection and exhibition of art has fed through the ranks, leading to some of the most important and recognisable artists of our time coming out of that country: the father of contemporary art himself Joseph Beuys, Anselm Kiefer, Georg Baselitz, Martin Kippenberger, Sigmar Polke, Neo Rauch, Gerhard Richter, Candida Höfer, Andreas Gursky, Rosemarie Trockel, Thomas Ruff, Thomas Schütte, Blinky Palermo, Katharina Fritsch, Franz Ackermann, Thomas Demand… The list goes on and on. Everywhere you go in Germany, every city, every town has a public art building and every one of them houses examples by nearly every one of these artists, bought at the beginning of and throughout their careers. And, because of this, Germany has created a school of influence like no other.

By cultivating a collecting culture at home that is nurtured by the state and maximises the exposure of its artists, Germany has created a story around the production of contemporary art, shaping the landscape for home-grown artists and generating buy-in from the rest of the world, therefore making it a significant global player. As part of this strategy, major exhibitions and surveys of contemporary art have been realised in the country, in particular DOCUMENTA (which takes place every five years in Kassel) and Skulptur Projekte Münster (which is held every ten years). These are now two of the most respected and renowned events on the calendar of the art world.

In order to coincide with DOCUMENTA 12, Skulptur Projekte Münster 07 and the fifty-second Venice Biennale, a 2007 exhibition entitled 'Made in Germany: Young Contemporary Art from Germany' took place in conjunction with three prominent institutions for contemporary and modern art in Hanover – the Sprengel Museum Hannover, the kestnergesellschaft and the Kunstverein Hannover. With fifty participating German and international contemporary artists, it highlighted the role of the state in building a formidable and flourishing artistic community in the country.

> *Specific German cultural policies and federal structures are involved here, which together facilitate and shape Germany's regional art scenes through the outstanding density, variety and continuity of art institutions, art colleges and funding schemes in the context of this country—be they at the traditionally strong centers of Cologne, Düsseldorf and Munich, or in the context of 'newcomers' like Dresden or Leipzig. Further, the intense institutional range of museums, Kunsthallen, Kunstvereine, as well as academies have contributed to a rise in qualitative artistic energy in the greater areas of Hamburg, Frankfurt and Stuttgart. Without doubt, the early presence of artist exchange grants such as the renowned* DAAD *(German Academic Exchange Service) in Berlin, as well as the academies with their professorships for appointed international artists, have unleashed lively international discourse on the art scene, which then became an essential component in the understanding of the national art scene.*[1]

What Germany has essentially done is create a brand. It has identified a distinctive thread in contemporary art practice that has distinguished itself, set its artists apart and led to a thriving art scene. It involves more than investing capital; it means crafting a concept to go alongside it, like you would for any consumer product. German contemporary art is weighty, laudable, impressive and museum-worthy – because this is the belief that has been constructed. By raising the profile of its artists at home, Germany has given them a platform from which to infiltrate globally, situating German contemporary art in the history books of the future, influencing art-making and the art market worldwide.

The notion of branding in the art world is not new and is not unique to Germany. Movements throughout art history have always had certain characteristics and personalities that adhered to their story and created community. Dada, Cubism, Abstract Expressionism, Pop – all of these movements had a kind of lifestyle that went with them. (Andy Warhol was probably the first to fabricate this with any serious commercial intent.) Likewise with movements that came out of certain schools or academies, such as Bauhaus; there was an infrastructure for support but, most important, there was camaraderie.

More recently, one of the most explicit examples of branding in the art world is the Young British Artists (YBA) movement of the 1990s. Again, the artists of this movement have become some of the most celebrated (in the literal sense) artists of now. Led by the artist (more significantly, the skilled marketer) Damien Hirst, this group, including the likes of Sarah Lucas, Gillian Wearing, Sam Taylor-Wood, Gary Hume, Mat Collishaw and the late Angus Fairhurst, studied together at Goldsmiths, London in the late 1980s under a teaching staff that included Michael Craig-Martin and Mark Wallinger. While studying at Goldsmiths, Hirst organised

'Freeze', an exhibition comprising the work of sixteen of his peers at university and the first artist-run exhibition of its kind, setting the precedent for the practice of artist-as-curator. This was the beginning of the YBA movement, which would later grow to include the eminent Tracey Emin, Marc Quinn, Tacita Dean, Rachel Whiteread, Gavin Turk, Chris Ofili, and Jake and Dinos Chapman amongst others, drastically changing the face of British contemporary art.

The YBAs were strongly supported by Charles Saatchi – founder of Saatchi & Saatchi advertising agency, major modern and contemporary art collector and something of an art institution in his own right. He collected their work ferociously and was their main sponsor and patron, nurturing and promoting the group and exhibiting their work in a series of shows at his St John's Wood space (where he had previously exhibited major American figures like Alex Katz, Richard Serra, Donald Judd and Cy Twombly, and some of the big German names like Kiefer, Polke and Richter). These YBA exhibitions garnered huge media attention, more for their brash aesthetic and controversial subject matter than the art itself.

Hirst and Saatchi took a fairly sensationalist approach to the development of the YBA profile and the most contentious exhibition of all was named in that vein. 'Sensation', an exhibition of works from the Saatchi collection, took place at the Royal Academy, London, in 1997 and later travelled to New York, causing outrage in both cities for the supposedly offensive nature of works by Marcus Harvey and Chris Ofili respectively. This, however, only served to reinforce what was now a very strong YBA brand, emphasising their importance and influence as a faction of contemporary art.

The YBAs were young Brits with attitude who were doing it for themselves; they were fearless, provocative and in-your-face. This is what they were selling and this is what the art world bought. But this was *not* just a passing fad. Hirst and Saatchi had created something of a phenomenon in British art, fostering a cultural revival that filtered through the entire industry. Major new art galleries and dealers were established during this period on the back of the YBA movement – Sadie Coles HQ, Victoria Miro, Jay Jopling's White Cube, Maureen Paley's Interim Art – and these remain prominent today. Many YBA artists were nominees and recipients of the prestigious Turner Prize and all have been collected by major museums around the world. And the rest, as they say, is art history.

So what principles here can be applied to the future of art in Ireland? How do we create a vision for tomorrow?

The crucial components for a successful matrix of art historical impact are talent, education, exhibition, exposure and collection. In order to perpetuate the future of art it is essential to break away from any parochial constraints and to be given a world stage. With the absence of a significant benefactor to provide a platform for institution-like exhibition, some strategic public programming is essential. The talent here exists (in fact, problematic societal circumstances have always been a catalyst for great art-making) but the market is small and private galleries are struggling to survive. This has led to a surge in underground activity. Artists are employing their entrepreneurial characteristics; artist-run spaces and exhibitions are contributing to a burgeoning scene at that level. However, without some unity and identity, it's hard to see what impact it may have.

What would happen if we combined these forces? What would happen if we took the energy, ambition and innovation of our artists and put a national stamp on it, amalgamating it all under one metaphorical roof? The Irish Museum of Modern Art could become a mobile body, occupying vacant spaces and buildings around the country, running public exhibitions that give weight and importance to the work of our artists, bringing young curators to the fore and defining a moment for contemporary art in Ireland. What if we created a brand for Irish contemporary art under this umbrella? We could build a narrative that could open up Irish art globally and see us into the future.

In Finland, we saw this come out of the Helsinki School. A country comparable in population size to Ireland and a city comparable in population size to Dublin have created a contemporary art movement in photography and have become known for cool, conceptual photography internationally. It could be said that this move mirrors the transition of the country itself from a somewhat isolated to an integrated place. It may be no surprise that the Helsinki School, one of the most recognisable Finnish brands (let alone contemporary art brands) today, was born in a city where the future-orientated Aalto University combines the higher education of art, technology and business. Beginning at a national level, the Helsinki School brand (which incorporates a permanent exhibition space called Gallery TAIK) has fed into the development of the entire Finnish contemporary art scene, irrelevant of media, and has influenced photography practice worldwide.

There has always been something very special about Ireland as a country, a reach and positioning that outweighs its size, and this is something that should be harnessed, nurtured and encouraged across the board. With the development of a national programme that cultivates a brand and identity for contemporary visual art practice in Ireland, there is no reason why Irish artists could not make a strong and significant impact on the global art scene. This begins at home, of course. It involves the provision of a framework in which emerging Irish artists can be elevated and celebrated. It means creating a wider context in which they can exist on an international level. It means promoting them, valuing them and advocating a discourse that is worthy of art historical inclusion. By defining a moment in time that is an expression of our culture and humanity, and in creating an identity with vision and foresight, we can give the credence and substance to contemporary Irish art that bestows the power to influence on a worldwide scale.

Notes

1 <http://www.artfairsinternational.com/?p=257>.

Chapter 4

John Egan

Banking 1

Having sold his first company at the age of 20, it seems John decided to make it his one-man-mission to encourage and bolster entrepreneurship in Ireland – and he's extremely enthusiastic about it. His credentials include: Dublin curator of the World Economic Forum Global Shapers; Irish ambassador to iCUE; co-ambassador of Sandbox in Dublin and CEO and founder of Archipelago (one of the largest communities of young entrepreneurs in Europe) and the 'Archie Talks' series. His preferred area of expertise these days, however, is on the future of banking. It's a subject on which he takes a long-term view and it's no surprise that banks around the world regularly consult him about it.

> *With something so important, a deeper mystery seems only decent.*
>
> John Kenneth Galbraith

For some time, discussion on the future of Irish banking has, unfortunately, been dominated by legislative reform and remunerative adjustments; a laboured debate on whether the banking industry is a broken arm of civic governance or the great evil of our time. Bankers have been castigated for being sociopathic megalomaniacs, government for being recklessly antisocial and we the citizens for being needlessly irresponsible. But the problems with banking in Ireland have prevented us from discussing what really matters: the future structure of the industry, its drivers and the inevitable influence of technology on the way we interact with finance and each other.

Given the public perception of the financial sector, it is important to distinguish between banks (the institutions) and banking (the industry). Banking is broken but it is necessary to reduce friction in capital markets. Banks, on the other hand, are organisations struggling to cope gracefully with the burdensome inevitability of their own demise. It is becoming increasingly likely that banks in their current form will cease to exist in twenty years. Retail banks are almost functionally obsolete. The traditional bank functions of savings, deposits, investments, security, trade, advice and financial management are almost all being fulfilled more efficiently by existing companies in peripheral industries. These companies have the wherewithal to execute traditional banking services more effectively and deploy them in a manner that is significantly more cognisant and in tune with the lives and needs of their customers.

Banks have long acted as proxies for the banking industry, so much so that it has become difficult to establish where one ends and the other begins. For many, 'banks' encapsulate or even define 'banking'. But banks are just agents, providing the seven services the industry offers, namely: security, lending, deposits, advisory, investment, trade and distribution of currency. Their monopoly on their provision of these services has been dictated by their ability to accept deposits. Banks, therefore, rely on two things to sustain their existence: (1) the circulation of physical currency and (2) customer trust in the institutions within the banking sector.

As long as governments could ensure (through regulation) that banks were operating in the best interests of consumers, customers had no reason not to trust banks and no reason to find alternatives. The combination of customer-focused regulation and standard free market practices would ensure that anyone operating in opposition to market demands would be eradicated. But the synthetic injection of capital into the banking industry since 2008 has created a cadre of government-protected oligopolists within the adjacently free market.

What has become clear is that there was an expectancy on the consumer side that a bank would check financial products on their behalf and only sell what it was appropriate for them to buy. At first glance, this may seem contrary to rational retail practice and philosophy – let the buyer beware, etc. But we have seen over time that in areas where expertise is required, risk is high or ambiguity is prevalent, there is a premium paid for ethical behaviour. We don't expect a bank to deliberately mis-sell us the wrong product because it would be ethically reprehensible to do so, much like we don't expect a doctor to mis-prescribe us medicine intentionally. Yet this is what we saw happen consistently throughout the boom years: the deliberate mis-selling of products to misinformed clients who believed the banks were working on their behalf.

The upshot being that banks like Anglo and Nationwide who propagate poor practice, mis-selling and antiquated business structures and strategies are artificially sustained by finance they have not earned. This has undermined consumer trust in the ability of the government to effectively police banks and their activities and, consequently, undermined consumer trust in deposit-based banking.

Traditionally, the response of the market to inefficiency is abandonment, boycott and the creation of alternatives – and that's exactly what we've seen internationally over the past eighteen months with the emergence of a number of bank transfer programmes, in particular Bank Transfer Day, MoveOn.org and Occupy. In the period 29 September–1 November 2011, CNNMoney reported that an estimated 650,000 American customers moved approximately $4.5 billion out of banks and into credit unions in what I would suggest is a minor indicator of the free market's Plan B. It hasn't happened to the same extent in Ireland, but it seems inevitable that it will happen over the next decade.

That change will see the most significant shift in corporate capacity since the Second World War. Stalwarts of the old guard will fall by the wayside, too inflexible to evolve, too scared to try. New players and ideas will emerge, many from the startup community but mostly from adjacent industries. Those banks who survive will be entirely different entities. We will witness the rise of the Networked Bank, the rise of the platform, the rise of the user and ultimately, the accelerated demise of the traditional notion of banks and banking – all thanks to the rapid evolution and deployment of new technologies and the internet.

The internet is now making redundant many of the traditional market barriers to banking. It is making it easier for independent, unregulated companies to play an increasingly significant role in all of the seven services of the banking industry. Over the last decade, we have seen the emergence of crowd-based reviews, suggestions and funding. It was only a matter of time before lending became part of that cohort. Peer-to-peer lending is a term referring to community-based lending initiatives like Prosper.com and LendingClub.com. These sites facilitate inter-member lending and borrowing at rates similar to – and often better than – banks, operating as a type of broker. Financial advisory sites like Mint.com, Buxfer or Geezeo offer users significantly more comprehensive advice than a bank advisor can, in a manner that's far easier for customers to use. Payment mechanisms such as PayPal, Stripe (founded by Limerick brothers Patrick and John Collison), Google Checkout and most recently Visa's new V.me platform, make it stunningly easy to buy things online. The list goes on. Each of the seven services of banking is being more effectively served or will be more effectively served in the next decade by bank alternatives.

The decline of the neighbourhood retail bank will give rise to a new world order of banking protagonists. Technology will fill the experiential vacuum between banks; deposits legislation will be circumvented and networks will dictate industry winners and losers. The current protagonists of the banking sector are behemoths whose huge power stems from industry barriers to entry, permission-based access to credit and personal deposits – characteristics that the internet is systematically sidestepping. Peer-to-Peer lending, crowd funding, digital wallets and currencies, payments technology and networked banking will redefine an antiquated industry.

This new world order will emerge within the banking sector over the next two decades, pioneered by a technologically-enabled cohort of leaders made up of the more dynamic of the old world banks, adjacent industries and startups. Banking will be mobile, money will be digital, branches will be exceptions and deposits will be an unnecessary balance sheet risk. Banking will be diversified, products will be customisable, service will be reviewable, advice will be crowd-sourced and ultimately banks will be *user-driven*, *open* and *networked*. This is the future of banking.

The cornerstone of this future is the Networked Bank. The Networked Bank is a platform-based bank focused on users rather than customers. It's a conceptual banking structure where individual banks act as intermediaries, facilitating capital flow between its users. Instead of accepting deposits and giving loans, the Networked Bank focuses on co-ordinating loans between their enormous user-base. Instead of depositing your earnings, the bank breaks up your funds into hundreds of thousands of units and co-ordinates a series of loans based on your own risk appetite, resulting in varying rates of return, completely defined by you the user. You get to create your own investment products and the bank earns an intermediary fee. The enormous user-base means portfolio diversification is almost perfect and risk, return and price can be predicted with extremely high accuracy. In short, it relies on data collection to reduce friction, improve liquidity and be the most efficient bank ever created.

But there are other companies out there, in non-banking-related spaces, who have become particularly adept at collecting and utilising data. When I shop online, Amazon recommends books that I might like based on previous purchases; Google filters my search based on geography and preferences; iTunes on what similar users bought; Facebook on who I interact with most. All of these companies use data they collect to improve my experience. The platform is subject to constant improvement and becomes increasingly predictive through user interaction.

These companies can make incredibly accurate predictions about our lives through aggregating and testing data on how we use their platforms. Facebook knows who is having an affair and when someone will leave their job. Google can predict elections and the success of products and movies before they're even launched. They can do this because they have extremely large and broad user-bases, often created and developed through social mechanisms, searching for and interacting with each other through content and transactions. They can also make accurate predictions on how to best deliver their own products. They can deliver what users want in a way they want, when they want. What if they used these same principles to build banking platforms? What would it look like if the world's largest technology companies set up banks?

Technology companies have access to all sorts of information: what we listen to, our technical capabilities, indicative metrics on our net worth, how far we travel every year, what we want people to know about us and our deepest secrets. Indeed, technology companies collect so much data about us as users that their banks would be able to make exceptionally accurate personalised financial recommendations over a significant time period.

But these technology-focused banks can do more than make accurate recommendations to us on financial products. They can help us to determine the efficiency of every financial decision we make, anywhere, anytime. They can help us to establish how we live our lives with a level of financial control, foresight and prudence that we couldn't access any other way.

By taking some statistically significant variables into account (e.g. the user's personal level of debt, account balance and monthly salary growth) and overlaying that with supplemental information (e.g. our music tastes, media consumption, internet usage and app purchases), the Networked Bank can make accurate recommendations on everything from birthday presents to how many children you should have.

By establishing some basic financial parameters and comparing them with tastes, age and productivity metrics, a company like Google can predict how much you the user are likely to earn over a month, a year and even over your lifetime.

Take Joe for instance. Joe is 32, in a relationship (with Jane) and earning €50,000 a year in the financial services industry. The Networked Bank can see from Joe's income that Joe's salary has only increased in line with the rate of inflation over the last six years and he has been paid by the same employer for that period of time. It is apparent that Joe is not being fast-tracked. Joe is not in demand; Joe is average.

The bank knows what industry Joe works in because he volunteered this information to Google/Facebook/Apple/Flipboard/LinkedIn/Twitter/Pintrest when he set up his account somewhere just after he got the job. But even if he hadn't, Joe owns a smartphone and Apple or Google can see from the map function that Joe travels to the financial district at 8 a.m. every weekday morning. The bank can then establish from some fairly simple algorithms what Average Joe's lifetime salary is likely to be. They can establish the relevant predictive variables (i.e. the music and apps) and cross-section that with macro-economic data, employment trends and search metrics. They can see how much debt he maintains and how fiscally responsible he is. Through social functions, they may even be able to tell if he's from a wealthy family and if there's access to alternative incomes. From all this information, the Networked Bank can create a completely bespoke financial management and purchasing schedule for Joe – for free.

This tool is a paradigm obliterator. This simple, free-to-access financial planner will be the greatest financial tool customers have ever had and it cannot be imitated by the banks. It can, for all intents and purposes, make all our financial decisions. For example, this bank will be able to tell Joe what groceries to buy and in what shops based on preferences, proximity, traffic and weather. It can tell him where the special offers are on the things he usually likes. If he lives in a high-crime area it can tell him the safest time at which he should go shopping, if there are any police around and even the likelihood of there being an incident en route. It can sync with public transport systems and shop stock levels and identify how busy the shop will be when Joe gets there. It can even help Joe stay healthy by integrating health apps that generate alternative, reduced-calorie shopping lists and take the next closest option so he can get a walk in.

But the Networked Bank can do even more. It can integrate Joe's and Jane's accounts and help them to choose affordable holidays that are within their combined budget, suggest date nights and anniversary presents and even recommend the appropriate time for them to get married and the appropriate amount of money they should spend on their wedding. It can also establish how many children Joe and Jane can afford to have, given the cost of raising a child in the city or country where they live.

The Networked Bank could be your life coach, mentor, personal trainer, teacher – so much more than a bank. And it's inevitable. If conventional banks stubbornly refuse to change or indeed stagnate to the point where change is no longer feasible, this is the future that will replace them.

Granted, there are serious privacy concerns that will need to be mitigated, but humans have long demonstrated an indomitable appetite for the shortest route from A to B and if any large technology company can create a product that makes our lives as users comprehensively more efficient, adoption will be absolute. Somewhere in Cupertino or Mountain View, California there sits a server with enough raw data on users for the likes of Apple or Google to create a bank that fundamentally changes the way we live.

This is the future of banking. It is not particular to Ireland, but for Ireland it is of particular interest.

Chapter 5

Orla Power

Banking II

Based in New York, Orla is an engagement manager in Ernst & Young's Financial Services Risk Management Advisory practice, having been Vice-President at JPMorgan Chase & Co. and, before that, Head of FSA Liquidity Risk at Santander UK plc. She obtained a BA in Business, Economics and Social Studies from Trinity College Dublin and went on to achieve her MSc Finance from the University of London. A Young Leader of the American Ireland Fund, she is an excellent role model for other young Irish women who aim to be part of the top echelons of finance on a global level.

My vision for Ireland is one where mothers and fathers proudly lean in to say, 'You know, my daughter works in banking.' That vision feels like a long way off at times. The Celtic Tiger years have left us with a terrible legacy. It is clear to us now that for many years in Ireland the wrong people were making the wrong decisions in banking. We are all paying for that now. It affects each of us in our daily lives, and so the hurt is deep. In the words of Maya Angelou, 'People will forget what you said, people will forget what you did, but people will never forget how you made them feel.'

The banking system, and people's trust in it, is broken. But if we are to work our way out of this problem, we cannot stay as disconnected as we are right now. We need to work our way out of this together. There are so many people in banking who are driven towards having a positive impact on Ireland. They want to stand tall and make a meaningful contribution to their local community – the community of their parents, family and friends.

In Ireland we have an impressive grasp of complex banking issues. We can talk all day about debt restructuring and promissory notes. We even discuss interest rates in terms of basis points rather than percentage points. But in our in willingness to discuss these esoteric issues, we ignore one crucial fact: banking, like any other business, is really about people.

Banking is deeply intertwined with our way of living in this world. The big events in our lives are shaped by banking and, because of this, we used to feel a connection to our banking system. We used to have a sense of ownership. We used to feel like banking was a collaborative industry. We used to think of banking as a chain that went all the way from High Street to Wall Street. I know this connection used to be there, because I see it in my own family history.

When I think of banking, I think of my grandfather. Granddad's workshop was a huge dark, dusty annex behind the main house, with an enormous wooden bench and bulky cutting machines on every surface. As children, we were on permanent warning that the workshop was 'very dangerous'. But of course it represented something different for my grandfather. As a young man in the 1930s, he had moved from Cork to London to work as a carpenter with Harland & Wolff, the company that built the *Titanic*. During that time, he transitioned from apprentice to journeyman, before ultimately producing his masterpiece and earning the accolade of Master Tradesman.

Professional accomplishment was enticing, but for Joseph Power romance beckoned and he returned home to Ireland to marry my beautiful grandmother, Mabel, with plans to put his skills to work for the people of Cork city. The stars aligned on his return to Cork, as a local businessman heard of my grandfather's skills and commissioned him to rally a team for the development of his personal home. While this was an incredible opportunity, my grandfather had no money upfront to pay labourers' wages. And so the branch manager at South Mall in Cork reached out and offered a bank loan for all of the money my grandfather needed to get the project off the ground. This meant that my grandfather could hire four men and get started on building the local businessman's home.

This first house-building project marked the start of a lifelong career building homes in Cork city. Years later when I visited Cork with my parents, we would walk around those neighbourhoods that my grandfather had helped to build. I have wonderful memories of wandering through St Finbarr's Park to feed the swans at the lough. My grandparents chose their personal home from one of my grandfather's neighbourhoods called Westbourne Park, named after the London area where he had lived in his twenties.

The combination of that first house-building commission and the willingness of the local branch manager to support the project made my grandfather's dream a reality. The bank and my grandfather joined forces to facilitate the transition from that first commission into a successful company. The local bank manager understood that it was my grandfather taking the ultimate personal risk. It was clear that my grandfather had all the necessary vision and skill, but he needed financial support if the project was going to work. In truth, each man needed the other in order to grow his business. And so the bank manager and my grandfather forged a business relationship that was loyal, supportive and solid.

When I think of banking, I think of my grandfather – and I also think of my parents. Both of my parents are retired teachers: my mother was a primary school principal and my father a secondary school teacher of maths and science. When they were starting out in life they had big dreams of buying their first home together. Back in those days the banks had, in the words of my mother, 'far stingier lending practices' so my parents had to make major sacrifices to get their deposit together. Twelve months later, with the help of their local bank manager, they found themselves with keys in hand, walking up the path to their beautiful, white, four-bed house with Spanish-style shutters on all of the windows.

It hadn't been easy for them. Even with all the scrimping and saving, even with all the stingy lending limits of the banks, it was still a scary financial leap. But they had a good relationship with the staff in their local bank. They knew that if things got rough, they could always pick up the phone and talk to somebody who knew how the whole thing worked. For my parents, the local bank professional was like the local family doctor: knowledgeable, trustworthy and always ready to listen.

My parents were delighted with their new home. They celebrated this major life event by sitting together on one of the cardboard boxes and sharing a Toblerone bar. Of course, this domestic calm was soon shattered once they had children! We arrived one by one and made our mark on the family home. Life moves along and the family events come and go. And somehow banking is intertwined with it all.

When my eldest sister, Denise, grew up and decided she was off to South America for a year, the experience was bittersweet for my parents. They were thrilled at her sense of independence and they savoured the time they had with her before she headed off on her adventure. But they also did a lot of worrying. What would she do if she ran into any trouble so far away from home? What would happen if she was stuck for money somewhere? My parents decided to go with Denise to the local bank and to talk about how money might be transferred if it was needed in a hurry. The branch assistant was around the age of my parents and she understood their worries. She arranged for a loan that Denise could use at a moment's notice if anything were to happen. Ultimately, it wasn't about my sister having extra cash in her pocket. It was about making my parents feel a little bit better waving Denise off at the airport. Yes, there were tears. But rather than tears of anxiety, these were tears of support and excitement about the fantastic adventure awaiting my big sister.

And then there's my sister Audrey, who works in the fraud protection side of banking. Audrey and her colleagues in the bank are great for local activities and fundraisers. I remember one time when they helped to redecorate a city centre nursing home. I know that's not the image that comes to mind when we think of banking, but it's true. There are plenty of good people out there. And there are plenty of good people who work in banking.

Therein lies the whole point: banking doesn't function without *people.* A computer model that measures risk is only as smart as the human who designed it. So as much as banking is electronic and automated, it also has to have human input. When we remove human interaction, we also remove human judgment and human accountability. We know all too well the dangers inherent in faceless, nameless people in distant boardrooms taking risks that affect other people's everyday lives.

We live in a globalised world and, of course, in many ways banking has to become more globalised. But as much as banking goes global, it also has to stay local. In our race to embrace all the wonderful new technologies and systems and efficiencies, let's not turn our backs on all the things that really matter on a local level, such as lending to small businesses. There is a thriving entrepreneurial and small-business culture in Ireland. Innovation and smart decision-making is at our core. But right now, the stats do not defend the rhetoric that Ireland is 'open for business'. During the period of January 2012 through December 2012, lending to non-financial corporations was down 4 per cent. While organisations such as Enterprise Ireland and Social Entrepreneurs Ireland are making incredible strides to support Irish people looking to make an impact at home and abroad, and to bring ambitious solutions to our environmental and social problems, banks too need to step up and lend to individuals and businesses when they need it.

The transformation needed in banking doesn't just involve the outwardly interactions between bank and customer.

Within banking, we could do with some changes. For a start, we could do with a lot more diversity. Irish banks need to aggressively pursue gender-diversity targets at both senior management and board of director level. Men alone are not to blame for the financial crisis. However, when decision-making power resides in the hands of groups that are thinking and acting in one cohesive manner, meaningful reviews of activities, along with any challenges to them, are limited. Diversity facilitates substantial debate about shareholder, customer and employee interests. Diversity can generate an environment characterised by bold debate and balanced judgment. Forcing that balance at senior management and board level attracts a strong pipeline of employees who aspire to future leadership positions; it also appeals to investors who appreciate the values of review, challenge and debate.

A lot of people who work in banking today feel demoralised and alienated. That is no good for Ireland's future or for the future of banking. The banking industry needs to show integrity from the inside out. Banks should have an accountable, responsible and courageous culture at their core.

Banking needs to find a new way of doing things. We need equality and honesty and fairness all the way from the branch to the boardroom. We need new thinking for a new banking system. I see evidence of this new thinking in the industry and that makes me hopeful. In fact, I have a feeling that all of this new thinking might actually lead us back to some old values.

Chapter 6

Padraig Mannion

Careers

Padraig is the founder of recruitment startup Superbly.co, which he describes as 'a virtual career matchmaker'. Both a founding member of Sandbox and a WEF global shaper, he has spanned the realms of creativity and business since he graduated from the Dun Laoghaire Institute of Art, Design & Technology in 2006 with a BA in Design Communication. He is a seasoned entrepreneur: he set up Studio Rua, a digital media agency, straight out of college and led it successfully for five years before this latest venture eventually took up all of his time. He is now focused on the future of careers and empowering Generation Y.

Like most people in Ireland, my early understanding of work was shaped by my experiences as a child. While growing up, there were a few solid truths: look both ways when crossing the road; always eat your vegetables; and if you want a steady career, get a government job. Be it as a teacher, a nurse, a guard or a postman, once you had secured that job, you could relax and enjoy your life and there would be a steady source of income that you could rely on until you retired. Get married, buy your house and start planning the kids – you were sorted. When you heard stories of people leaving government jobs it was always said in the context of shock. They were seen to be foolishly throwing away the great state pension.

The same stability that came with a government job was also perceived in large corporations and multinationals. The general understanding was that you measured the quality of the job by the length of time you could remain in it. This was a stable career – and a career was all about stability.

With this stability came a form of institutionalisation. Companies so intent on keeping their staff for life took on the responsibility of training them as they rose up in the organisation. This linear progression was also a rudimentary form of career management.

The disadvantage of this career progression strategy was that companies became top-heavy, as the staff moved along the organisation conveyer belt of promotions and wage increases. Examples of this can still be seen in some public bodies and large corporations today. It is a sure recipe for many problems, primarily the recurring cost of high-level staff, the lack of true leadership among management and, therefore, the resulting inability to react effectively to market dynamics.

To combat this, companies gave more focus and strategy to their career management practices. The resulting experience for employees was one of greater success within the company and a sense that the company had a plan for them, but it was still one of passive engagement. A boss decided what was best for an employee and the employee had very little visibility of how their organisation was structured or where the opportunities lay for them. But this was not a terrible thing; they were still able to rely on the perceived stability of their job.

Fast-forward to the marketplace of the 1990s. Years of corporate restructurings, mergers, acquisitions and downsizings resulted in workplace trends characterised by job insecurity, flatter organisations and fewer promotions.[1] Corporate executives of this era were much more savvy when it came to managing their staff, but they were also conscious of the extra costs that came with managing all these careers. The HR trend moved to shift accountability for career management from the employer to the employee, offering formal interventions such as training to help employees learn to take greater responsibility for their own careers.[2] It is at this stage that corporations really drew a line in the sand and made the clear statement that, although they were willing to support and accommodate career management, the success of each employee within the company was the responsibility of the employee themselves.

Which brings us to today. In Ireland, the government has an embargo on public sector recruitment; those who have a public sector job are feeling the pinch of austerity, while their pensions aren't looking like the nest egg they once were. Corporations now expect individuals to manage their own career path, and although they provide greater internal visibility, the large majority still expect employees to be able to navigate their own way to the top. The recent economic crisis has made the jobs marketplace extremely competitive. And with so many young people relocating to places like Canada and Australia each year, mass emigration is well and truly back on the agenda.

So how is Generation Y (Ireland's 14- to 27-year-olds) tackling this situation? How are they equipped to deal with challenges in their own careers?

Let's first look at a brief overview of Gen Y and some of the characteristics that define them in the workplace. Gen Y has been tagged as self-entitled, overly-ambitious dreamers who don't want to pay their dues and are only concerned about higher pay and more time off. On a more positive note, however, they are also described as future-oriented, ready to contribute now and opportunity-driven. They remain admirably optimistic in the midst of the current economic turmoil and they are highly restless. They seek to earn greater opportunities for rapid advancement and more responsibilities at a younger age.[3]

These attributes, positive and negative, can be seen in the increasing amount of young graduates creating their own startups straight out of college, as well as excelling in established companies. It can also be seen in the numbers of young people emigrating from Ireland every year. The characteristics of both of these groups – those who stay in Ireland and those who don't – are perhaps perceived differently. The young entrepreneurs and intrapreneurs who stay are celebrated for their bravery and independence, while the ones who choose to independently pursue their careers in a foreign land are mourned in the national press.

This opinion might seem logical to some but, to me, what these two groups of people are doing is actually very similar. They are both taking massive personal action to achieve the career that they desire for themselves. And if we revert to the career management trends of corporations over the last few decades, we can see that this is exactly the ideal that has been encouraged: personal career management.

The focus of the modern career is now on opportunities to advance towards ambitious goals, with value put on meaningful development. It is this search of meaning in work that is often confused for arrogance by older generations. In times gone by, people were privileged if they had a career at all, and it was disrespectful to leave one behind because you didn't feel like your work had enough 'meaning'. That is, however, exactly what we are seeing today. Where previously a CV with more than three jobs over a ten-year period looked suspicious, it would now be quite normal, perhaps even expected.

Gen Y embraces mentoring from more senior staff members and prefers to work in teams. As such, they are very suited to take personal responsibility for their own careers. Companies are therefore under pressure to supply the tools that allow their Gen Y staffers to do just that, while also satisfying their constant desire for feedback and rewards.

As Gen Y is always on the look-out for more opportunities to advance their careers, there is a rapid embrace of new technology that lets them do so easily. While their predecessors, Gen X, are firmly situated on career websites like LinkedIn, which profile themselves and their achievements, more Gen Y-suited services like Branchout and Silp are now springing up on Facebook and attracting large numbers. These tools allow people to discover career opportunities through their friends and facilitate introductions into organisations that they want to work in. They are perfect examples of how Gen Y is leveraging its own technology to benefit career development. However, they still fall short of personal career management as an ideal.

Personal career management is defined as the degree to which one regularly gathers information and plans for career problem-solving and decision-making. It involves two main behaviours: one related to continuous improvement in one's current job (i.e. developmental feedback-seeking) and the other related to movement (i.e. job mobility preparedness).[4]

While services and support pertaining to such behaviours can often be found at a basic level in schools or colleges, there is a sudden halt when people enter the workforce. One growing trend in Ireland, however, does provide such information: the international job fairs that are increasingly held throughout the country.

I would encourage people not to see emigration as a negative thing or a path-of-no-return for the people who choose it. Ireland Inc. has benefited greatly from the Irish diaspora that extends around the world. Many people who left Ireland in the past now hold powerful and influential positions in successful companies from San Francisco to Shanghai. Not only are they now in a much better position to leverage Ireland on an international stage, but they can also provide amazing insights and mentoring for the young people of Gen Y.

If we truly aim to encourage and facilitate personal career management in Ireland, as I believe we should, then we need to let people explore their options until they find work that suits them. As we have noted, this is a trait that Gen Y will action whether we like it or not. Ireland is a small island nation with a population the same size as Greater Manchester, so if we are to advocate career exploration we will unavoidably encourage people to leave Ireland as well. I believe we should take advantage of such an inevitability and provide those who choose to find work in another country with as much support as possible.

To really embrace the future of Gen Y careers, we should remove the stigma of leaving Ireland, so that in turn we might reap the rewards of talent returning to Ireland.

Ireland has been particularly progressive in its recruitment of multinational companies to our shores and the (usually very successful) corporations who do come here rely on a steady stream of employees with the required education and skills. We have also been very successful in recruiting staff into the country to fill any gaps that inevitably exist within the native Irish workforce. Companies like Facebook and Google, with major headquarters in Dublin, rely heavily on international candidates to fill all their requirements and they do a brilliant job of promoting these opportunities abroad.

This is an amazing way of encouraging really talented people to move to Ireland. And what we are now seeing is that many who came to these shores originally to work for large multinationals have moved on professionally to pursue their own ventures but, crucially, they are doing so here. This development will obviously boost the Irish jobs market and it is a welcome cycle that could reap benefits for years to come. There are people of Gen Y in other countries who are more than willing to take their explorations to Ireland in order to find a career that is just right for them.

In the same way that people of Gen Y do not see one company as providing them with their entire career, they will also refrain from restricting their career to any one country. Ireland is in a unique position in that our Gen Y can avail of visas in many countries around the world, bringing with them highly valuable skills to enrich those economies. Crucially, however, they will also develop more skills and knowledge that could benefit their home country in the long-run.

Ireland has an opportunity to position itself on the international stage and to lead in personal career management, benefiting all companies hiring staff. Business in Ireland can gain excellent exposure, not only to the young, ambitious Irish workforce but also to an extensive international pool of Gen Y talent.

Notes

1 Hall, D.T. and Mirvis, P. (1995), 'Careers as Lifelong Learning', in A. Howard (ed.), *The Changing Nature of Work*. San Francisco: Jossey Bass.

2 Brockner, J. and Lee, R.J. (1995), 'Career Development in Downsizing Organizations: A Self-affirmation Analysis', in M. London (ed.), *Employees, Career and Job Creation: Developing Growth-oriented Human Resource Strategies & Programs*. San Francisco: Jossey Bass.

3 Analysis of the views of 860 Gen Y (ages 19–27) employees of Fortune 500 companies across twenty industry sectors who responded to an online questionnaire entitled 'New World, New Workforce'. The questionnaire was distributed by Zoomerang and Deloitte between 12 November 2008 and 7 January 2009. No Deloitte employees participated in Deloitte's 2009 survey entitled 'Generation Y: Powerhouse of the Global Economy'.

4 Ernst Kossek, E., Roberts, K., Fisher, S. and Demarr, B. (1998), 'Career Self-Management: A Quasi Experimental Assessment of the Effects of a Training Intervention', *Personnel Psychology*, 51/5, 935–60.

Chapter 7

Bella FitzPatrick

Crafts

Bella is project manager with The Undergraduate Awards. In this essay, she proposes a simple, fun idea that could be easily implemented and could have many benefits; she also unearths some interesting facts about knitting. Never stuck without a pair of needles or some wool, Bella founded Knit Soc (the knitting society) in Trinity College Dublin, where she obtained her BA *in Medicinal Chemistry in 2012. She is also the founder of Bowvember, an initiative that organises groups to knit and sell bows in aid of prostate cancer research during the month of November each year.*

In lectures, I could usually be found sitting front and centre and listening carefully. Yet I was used to professors stopping mid-sentence to give me a baffled look. You see, I was also usually knitting. I wasn't sure why, but I could pay more attention if my hands were busy with a pair of needles and a ball of wool. I even used different colours based on what topic I was doing. It was as if I was creating a physical manifestation of the knowledge I was absorbing; tangible evidence of what I had learned.

However, I only knew the basics of knitting and I wanted to get better. So my friend Hannah (a great knitter) and I decided that if there wasn't a place to go to learn more needle-craft skills, we'd have to make such a place ourselves. Thus, we set up a knitting society in our college. Before long, many amazing knitting enthusiasts came out of the woodwork. We started to host technique classes and to provide materials for beginners. Knit Soc, as it was known, was a friendly place to come and learn a craft and drink some tea in our weekly meetings, dubbed 'the stitch n' bitch'. Our motto was: *Knit Fast, Die Old!* We had around five hundred members in our first year; it was far more popular than we could possibly have imagined. Soon Hannah and I had to assemble a committee of ten people to help us run the society. Interestingly, from the very beginning Knit Soc maintained an even gender divide. Today it is managed by a committee with a male–female divide of around 50:50 and it has a male chairperson.

The society became so successful in its first year that one of the college newspapers wrote a small article about it. The article mentioned my habit of knitting during lectures. Knit Soc got plenty of attention as a result of the article. Unfortunately, not all of the attention was positive. A few months after the article appeared, I met with one of my chemistry professors to talk about the possibility of a summer research internship. I was surprised when, in the middle of the meeting, he stopped the conversation and said he had a bone to pick with me. He referred to the article and its comment about my habit of knitting during lectures. He said that the piece was very damaging to the chemistry department, since it gave the impression that I was neither a committed nor a hard-working student.

I was taken aback. Why would the chemistry department take a college newspaper article so seriously? I sincerely doubted that the department had ever lost any funding on the basis of articles such as these. Also, I knew I could knit *and* take in other things. I could knit in the cinema without missing any plot-twists. I could knit while having conversations and not lose any friends because of it. I knitted *all* the time and it had never caused any trouble. The act of knitting didn't impede me from paying attention – it actually *helped* me. I explained this to my professor, but he just rolled his eyes. I left, baffled but determined to find out more. I knew I benefited greatly from knitting. Was I the only one?

I felt misunderstood. You see, for me knitting wasn't a flippant thing at all. People joined Knit Soc for all sorts of reasons but for me it wasn't just for fun – it was a pretty serious matter. I needed to knit. I have a learning disability that affects my short-term memory. And in my college days, knitting really helped me. In the beginning, I didn't know *why* it helped me so much – I just knew it did. After my chat with the professor, I decided to do some research in defence of knitting. All of the statistics and all of the facts that I unearthed confirmed for me what I already knew to be true.

The truth is: knitting is good for you because knitting makes you happy. It releases serotonin, a neurochemical involved in appetite, sleep and mood. People suffering from depression have low levels of serotonin, but knitting (along with exercise and eating healthily) helps to naturally increase serotonin levels. Knitting also releases dopamine, the neurochemical involved in reward (which is usually released after eating, having sex or taking certain drugs). Dopamine is also associated with addiction and, as such, knitting has been found to help people kick habits such as smoking and over-eating. Knitting also helps people to relax. Brain scans show that people knitting are in a similar state to those meditating.

It's fair to say that knitting is an enjoyable activity. Although, that's hardly a surprising discovery. If it wasn't enjoyable, why would so many people do it at a time when professionally-knitted items can easily be purchased on Amazon? (It certainly isn't for the street cred.) But knitting has so much more to offer than pure enjoyment. Knitting really is good for your brain. Those who knit regularly are statistically less likely to develop dementia and Alzheimer's disease. Like puzzles or arithmetic, knitting has a stimulating effect on the brain. And repetitive visual-spatial tasks such as knitting have been seen to put people into what is referred to as 'theta rhythm'. This pattern of brain waves occurs when we are in a light sleep – not fully asleep and not fully awake – and it is linked with our most creative selves. Hence all those times when you wake up knowing you've just had a great idea. (Keeping a pen and paper beside the bed is a handy way to capture these thoughts, by the way.)

And there's more: knitting has been linked to improved concentration. It has been found that, once you become an experienced knitter, then knitting while talking or listening is not classified as multi-tasking. The knitting occupies the subconscious with an undemanding task, leaving the conscious mind more focused, ready to receive and process information.

Of course, I already knew these truths from my own experience. The effects of my learning disability subsided as soon as I first took up my knitting needles. My short-term memory disorder means that I have a tendency to store things in my short-term rather than my long-term memory. I have trouble remembering sequences (e.g. months of the year) and this affects how I interpret questions and information. When I first started college, I had to write hundreds of pages of notes, never miss a lecture and study hard in order to maintain my grades. I did this not to be at the top of the class but just to keep the information in my brain! I was unable to read something and retain it. I could only do so if I wrote it out over and over.

However, I soon realised that I could process more information during lectures in which I knitted. Noticing this, I started knitting while reading my textbooks – and that also helped me enormously. I could practically feel my synapses firing and neurological links being formed. Knitting, and teaching knitting in the stitch n' bitch sessions of Knit Soc, helped to decrease my stress levels as well. For me, knitting became so much more than a way to make thrifty Christmas presents.

I know from my own experience that knitting and other crafts can be amazing educational tools. I think that crafts should be placed just as high on the school curriculum as maths, languages, sports or anything else. Crafts are simple to teach, cheap to do and have so many benefits. I wish I had started knitting from a younger age – it might have meant less of a struggle in school.

Crafts can be used to tackle very heady concepts in the classroom. Crochet (very similar to knitting, but using just one needle with a hook) can be used to demonstrate some complex mathematical concepts. Crochet is similar to the binary number system on which computers operate, with a stitch representing '1' and a missing stitch representing '0'. In fact, early computer programs were based on equipment used to operate textile looms. Crochet has even been used in the past to best illustrate hyperbolic geometry. Daina Taimina, a mathematician from Latvia, used crotchet to create a strong, durable model of the hyperbolic plane, which is a concept used in Einstein's theory of relativity.

Crafts play a huge part in the Steiner Waldorf teaching method – a humanistic approach to education based on the educational philosophy of Rudolf Steiner. In Steiner Waldorf schools, knitting is taught to children as young as 6. In the Steiner Waldorf framework, it is believed that crafts allow children to assert a healthy degree of control. Crafts also build skills of attentiveness, which are so important for later life. In order to knit properly, a child must count stitches and rows; this means that numerical skills are reinforced in a challenging yet enjoyable manner, resulting in the child creating something they can then be praised for. Knitting has also been shown to help children with ADHD. And because there is an immediate reward and evidence of the effort put in, knitting is most popular among children with attention and learning disabilities and children who regularly misbehave.

If the evidence of the link between knitting and increased concentration is to be believed, then knitting is the perfect activity for schoolchildren: it is cheap, safe and easy, with a potential bounty of benefits. Higher levels of concentration in the classroom don't necessarily require the use of experimental drugs. Perhaps all that's required is some knitting. How about knitting a scarf, a hat or (for the advanced children) a pair of gloves?

Crafts can really affect what happens in the classroom. And what happens in the classroom affects us all, because children's experience of education shapes who they are. When children get an opportunity to create something, it affirms them as individuals. Each person needs to have their identity affirmed. It is wonderful if this sense of identity can be affirmed in the classroom. And sometimes it happens in an environment that is far removed from the classroom. Lynn Zwerling knows something about that. She is the co-founder of a group called Knitting Behind Bars – a knitting group in a prison in Maryland in the US. Zwerling has stated that the male inmates were reluctant to participate at first, but quickly started to enjoy knitting as an activity they could have control over and take pride in. One of the prisoners said: 'You have to watch what you're doing, otherwise your stitches will become loose or tight or you'll skip stitches. It almost makes you feel like you don't have to be anything. You're all sitting there knitting. You can just be yourself.'

Being free to be yourself is a powerful thing. It is only when a person has a strong sense of identity that they can give something back to society. The best education provides children with this sense of identity. It creates individuals who are able and willing to contribute. Crafts can help children to develop a strong sense of self and a strong sense of community.

Education should aim to make children more socially aware, but with this social awareness often comes a huge sense of powerlessness. Children don't have money to give and they can't volunteer. They feel burdened with knowing the injustices of the world and being unable to do anything about them. This is where crafts come in. Children can contribute to numerous charitable initiatives involving crafts: Penguin Sweaters are knitted by volunteers and provided to penguins in rehabilitation from the effects of oil spills; Project Linus accepts blankets for needy children; Innocent Smoothies have a hat-knitting campaign for Age Action Ireland; and there are several initiatives that require clothes to be made, e.g. for premature babies. When children have a chance to contribute in this way, everybody benefits. Worthy charities get much-needed donations and children learn how to be socially mindful.

Fundamentally, an emphasis on crafts is an emphasis on creativity. We all know the importance of creative outlets for children. Crafts help children to express themselves in an imaginative way. Learning crafts gives children the confidence to try new things.

Knitting and other crafts might seem very far removed from the problems of Irish society – but they're really not. Problems in an economy and problems in a society are solved by people. Ireland needs people who are creative thinkers. Ireland needs people who are problem solvers. So Ireland needs to provide for its children an education that fosters all of their inherent creativity.

But not all children (or adults for that matter) would consider themselves 'creative'. It's a word we usually reserve for the painters, the sculptors and the people with artistic talents. So young people who are oriented towards business or science often shy away from art classes or other activities that they believe are for 'creative people'. What a shame! Ireland needs creativity from *everyone*, including our future businesspeople and scientists. Crafts such as knitting have technical and practical aspects that everyone can master. This means that even the least artistic children can be creative.

Crafts take education out of the realm of rote learning and into the realm of creativity. Crafts are less about learning the established facts and more about creating vibrant solutions. The children of Ireland have so much creativity but it needs to be fostered now. After all, a stitch in time…

Crafts teach children how to build things little by little. Every knitted scarf starts with a single stitch. Through crafts, we can see that small incremental steps lead to big results. So I'm quite certain it's no accident that we call it 'the fabric of society'.

Further Reading

Doyon, J. and Benali, H. (2005), 'Reorganization and Plasticity in the Adult Brain During Learning of Motor Skills', *Current Opinion in Neurobiology*, 15/2, 161–7.

Duffy, K. (2007), 'Knitting Through Recovery One Stitch at a Time: Knitting as an Experiential Teaching Method for Affect Management in Group Therapy', *Journal of Groups in Addiction and Recovery*, 2, 67–83.

Fabrigoule, C. *et al.* (1995), 'Social and Leisure Activities and Risk of Dementia: a Prospective Longitudinal Study', *Journal of the American Geriatrics Society*, 43, 485–90.

Maceachren, E.J. (2001), 'Craftmaking: A Pedagogy for Environmental Awareness'. Unpublished PhD thesis, York University (Canada).

Riley, J., Corkhill, B. and Morris, C. (2013), 'The Benefits of Knitting for Personal and Social Wellbeing in Adulthood: Findings from an International Survey', *The British Journal of Occupational Therapy*, 76/2, 50–8.

Scarmeas, N. *et al.* (2001), 'Influence of Leisure Activity on the Incidence of Alzheimer's Disease', *Neurology*, 57/12, 2236–42.

Spurgeon, K. (2007), 'Knitting is Good For You' <http://kristinaspurgeon.myefolio.com/evidenceofcompetence2>

Taimina, D. (2009), *Crocheting Adventures with Hyperbolic Planes.* Oxfordshire: A K Peters/CRC Press.

<http://www.good.is/posts/prisoners-transform-through-knitting-behind-bars>.

<http://www.prlog.org/10179509-knitting-and-crochet-offer-long-term-health-benefits.html>.

Chapter 8

William Peat

Diaspora

William has been involved in researching the Irish diaspora for several years. In particular, he has researched versions of Irish identity within the San Francisco Bay Area, a topic on which he subsequently produced his MA thesis at NUI Maynooth. He has spoken at international conferences and acted as an advisor for several diasporal projects, including WorldIrish.com. I first met William through his NGen Ireland initiative, which he co-founded in 2012 as a new social venture aiming to discover, accelerate and celebrate Ireland's next generation of changemakers.

Migration is an integral part of the human experience; however, the Irish are unique in the popular knowledge of their migrated brethren. We are currently moving away from the one-dimensional view of migration as 'problem' and are developing an awareness of the many hues of migration stories. While we are still affected by the ghosts of our past, we are moving into an unprecedented period of communication and understanding around being 'Irish' in a global sense.

The following extract is from the old Irish song 'Three Leaves of Shamrock' and it is, to say the least, depressing. The purpose of this song is quite clear: to guilt a soon-to-be exile into sending back remittances to their family.

Three leaves of shamrock, the Irishman's shamrock,
From his own darling sister, her blessing, too, she gave;
'Take them to my brother, for I have no one other.
And these are the shamrocks from his dear old mother's grave.

'Tell him since he went away how bitter was our lot,
The landlord came one winter day and turned us from our cot;
Our troubles were so many, and our friends so very few.
And, brother, dear, our mother used to often sigh for you.

"Oh, darling son, come back!" she often used to say;
Alas one day she sickened, and soon was laid away.
Her grave I've water'd with my tears, that's where the flowers grew,
And, brother, dear, they're all I've got, and them I'll send to you.'

It was traditionally sung at an 'American wake', a peculiarly Irish event in which a community got together in order to hold a living funeral for the departing individual. 'Archaic in origin yet adapted to modern exigencies', according to Kerby Miller:

The American wakes both reflected and reinforced traditional communal attitudes toward emigration. Indeed, these rituals

> *seemed almost purposely designed to obscure the often mundane or ambiguous realities of emigration, to project communal sorrow and anger on the traditional English foe, to impress deep feelings of grief, guilt, and duty on the departing emigrants, and to send them forth as unhappy but faithful and vengeful 'exiles'.*[1]

This wake tells us two things about people's thought processes: firstly, that the individuals leaving were an economic necessity to their families back home, and secondly, that communicating with them once they left was sporadic at best. Sending someone off to a foreign land was a gamble, one with big risks and potentially big rewards. However, it is not the rationale behind them at the time that interests us here; it is the legacy that is important. The American wake as a propaganda campaign was extremely successful in that it created a deeply rooted image in the Irish national psyche about migration. Lurking around the deepest recesses of even the most globalised Irish person is the image of a coffin ship and the huddled masses vying for a spot in steerage, but for a variety of reasons this is now a fantasy akin to the bogeyman hiding under a child's bed.

It is true that not all migration stories are happy ones and that with current economic conditions more and more people find themselves having to seek employment abroad. However, it is important to remember that while they may be linked, current economic difficulties are not intrinsic to migration and vice versa; the simple fact is that moving abroad is less painful, less isolated, more familiar and easier than ever before and it will continue to become more normalised in our society into the future.

Transport and communication technologies, especially since the 1980s, have fundamentally changed the way we interact with the world and each other. Cheap, plentiful flights have brought the world much closer to us and there is now no major metropolitan area that is not physically accessible within 24 hours. The implication of this is that no one is truly far away, disconnected or unavailable and, therefore, if anything goes wrong they can come home or family members can travel out to help. Yes, this is overly simplistic – it doesn't take into account a plethora of factors – but ultimately the fact that this kind of travel is a possibility has significantly changed migrants' views on how they will face their journey and settlement.

The range of options and problems will shift from one in which travel and settlement were exceptionally hard but finding legitimate work was relatively simple, to a future of greater monitoring of people via the creation of a strict two-tiered system. The two-tiered system is one in which there are 'legitimate' and 'illegitimate' people living in a country or, in other words, those who have legal status in a country and those who do not. We already see this with the plight of the 'undocumented Irish' – and several million more people from other nationalities – in the US. However, the monitoring of this system is becoming stricter and more all-encompassing. Within the next decade those who conform to the 'system' will find it relatively easy to move abroad legitimately, but those who do not will find it much harder. The ability to digitally tag people entering and leaving a nation, along with digitally tracking their movement while there (through their electronic transactions, check-ins, etc.) will significantly increase. This can already easily be done using technology, so no human intervention is required to classify a person as 'legal' or 'illegal'; this greatly streamlines and vastly inflates the capacity of the system to vet people.

In the future, the Irish will have to take greater care to make sure they fit the categories of a legitimate person in order to make themselves able to travel and work in a plethora of countries. A good education, clean record and multilingual abilities are quickly becoming prerequisite criteria to be a desirable employee. Educationally speaking, Ireland has always gone to great efforts to conform to international standards. An interesting offshoot of our particular form of Catholicism is that in many countries the Irish are seen as educators because of the high volume of Irish nuns and priests who became teachers and school founders overseas. So as we move abroad, there should be an easy access to jobs. However, the lack of a second language is something we need to tackle very quickly amongst our young people. We are currently one of the five EU countries where respondents are least likely to be able to speak any foreign language (60 per cent). While this trend is declining as more of us speak a second language, it is still not seen as a basic skill in Ireland.

Ireland as an economy is one of the most open in the world, with Ireland ranking as the third 'most globalised nation' on earth.[2] With over 87,000 people leaving the state in 2012 and around 20,600 returning from living abroad, the Irish are truly a people on the move. We will do this more and more as time goes on and this will feature as a normal part of growing up because skills and industries will become more internationalised. If we look at the current relationship between the US and Ireland, we see that approximately 115,000 people are employed by US companies in Ireland but approximately 120,000 people are employed by Irish companies in the US. The development of this relationship will naturally lead to more and more people moving between two countries but within one company. This trend is being replicated in the UK, India and Asia-Pacific, along with the more traditional waves of individuals moving abroad. As we move away to other communities in the future, we will continue to communicate with each other in an intimate way with the use of technology. With over a billion people on Facebook, social media as the number one activity on the web and 69 per cent of parents being 'friends' with their children on social media sites, we are truly connected online. It is the intimacy of these connections that is so vastly different from previous generations.

The postal services and telephones have been around for generations, allowing for information to be passed between individuals on other continents. Nowadays people don't just receive reports: their social media provides them with up-to-date information on all aspects of their friends' and families' lives. The Facebook Data Centre now holds patterns of behaviour for when people are most likely to get into a relationship or come out of one. (Christmas is the best of times and the worst of times, apparently.)

When Apple launched their FaceTime programme, the ad conspicuously showed intimate family moments being shared through the iPhone; Apple were clearly portraying the notion that technology and family events can now go hand-in-hand. Grandparents can be at family events by being able to see and hear them on a plethora of electronic devices. While in no way replacing actual contact, technology has created a grey area of relationships in which we will be able to remain in close contact with people who are physically on the other side of the world. Advancements in travel have also made it possible to visit those people more often and in an easier manner than ever before. Certainly for friends and family, the world has never been so small.

This technological advancement will also change the way we interact with each other and with society at large. Robert Putnam's *Bowling Alone* popularised the concept that society is moving from a model based on communitarian values to one that privileges the individual. Social media (specifically multi-user sites such as Quora) has heralded the decline of traditional institutions being the bastions of knowledge, ringing in an era when individuals are able to access expert knowledge freely and easily. We will start to relate to one another on a much more individualistic basis than previously. In work, people will see themselves as a single commodity that has skill to be sold to employers, as opposed to being loyal and subservient units within a company. All of this will have a fundamental impact on how the Irish will relate to one another as they move abroad.

Previously, the Irish centres, county boards, parish halls and GAA clubs were key to survival for any newly arrived migrant. However, as we move forward, more of us will be 'citizens of the world': individuals with transferable skills that are respected and sought-after globally. Again, using technology, we will be able to look for employment, accommodation and even love online (one in five couples now meet online). We can access local knowledge anywhere and thus no longer need to join these community projects. Instead, our own personalities and interests will define who we interact with. The GAA is the exception that proves the rule as it goes from strength to strength. The decline of the county boards and the Irish centres has happened alongside the rise in popularity of GAA. Always a favourite of the Irish, the GAA has now become an indispensable part of the Irish community abroad. While dinner dances and committees seem passé nowadays, Gaelic football and hurling are still fun and so they have allowed the GAA to thrive. The GAA itself has also reacted to changing times and consistently proves itself to be one of the most ambitious and efficient organisations in Ireland.

Community-based organisations helped people settle into their adopted community, but they also demanded a certain amount too. Conforming to the group was an important element of success, but this is not so anymore. The experience of being in the diaspora will no longer be one in which people constantly refer back to their Irishness; it will become much more about finding local individuals who share interests and hobbies. Where those interests are to do with Ireland, then there can be overlap between Irishness and living, but the two will no longer be intrinsically linked. If you go to Silicon Valley, you are much more likely to find an Irish employee going to a networking event full of all nationalities working in the same industry, as opposed to a meeting in an Irish centre that is full of the same nationality working in all industries. With members of the diaspora intermingling with peers around the world, they will be gaining new knowledge and experience and this is easily going to become a major commodity to Ireland.

There is an appalling term used all too often nowadays: the 'brain drain'. Out of the 87,000 or so people who left Ireland in 2012, it is estimated that 39,700 were between the ages of 25 and 44. These, therefore, are the most productive, upwardly mobile, educated demographic in the state; vast numbers of this demographic are leaving. This has led to the fear that we are educating our population only for them to head abroad and give their skills to another country. However, since over 52,700 people moved to Ireland in 2012 it is obvious that there is a two-way exchange going on.

All of the major multinationals in Ireland have a significant number of educated foreign nationals working for them and they bring with them their education, work ethic and ethos. As much as the Irish are out in the world teaching, we are being taught at home by the world too. Contained within the 52,700 or so people are approximately 20,600 Irish people who returned home from living abroad. There is clearly an upward trend in the circuitous nature of this exchange. Kingsley Aikins, head of Diaspora Matters, has spent the past few years extolling 'brain circulation'. The premise of this term is that Irish people who go abroad will learn new skills, new practices and cultural outlooks. Those who return can enrich our own culture, alongside foreign nationals who come here and end up enriching us as well. This is certainly the case for those who do return and it will continue to be so in the future. However, it is also true that a lot of people who leave may not come back, but this is far from the end of their connection to Ireland.

Now more than ever the government and private organisations are reaching out and connecting the Irish around the world to one another and to Ireland. Connect Ireland, WorldIrish.com, NGen Ireland and Ireland XO are all home-based organisations reaching out to the Irish around the world in novel ways and creating connections between diaspora and the state like never before. WorldIrish.com bills itself as a space for 'Irish people and those with an affinity for Ireland together in one place'. They now have a membership of over 70,000 people from around the globe. Connect Ireland are reaching out to the Irish abroad with the simple idea of bringing jobs to Ireland and giving the tax back on each job sustained for more than a year. Ireland XO has decided that, rather than waiting on Irish-Americans to come searching for their roots, it would go and search on their behalf. (Once they find people with links to a particular place, they invite them over for a week of organised activities with locals, giving them a truly unique and authentic experience.) Abroad, there are the Irish networks such as the Irish Network USA, The Lansdowne Club Australia, The Farmleigh Fellowship Asia and the Irish Club Moscow. All of these organisations connect Irish individuals together in their particular location and provide great networking opportunities at local events and further afield through affiliations with a wider network of organisations. This interconnectivity will create a new, natural, Irish network that is global, online, personal and vested in individuals.

Overall, the Irish will be dealing with their intimate relations, their fellow countrymen at home, their fellow countrymen abroad and new people abroad in a cacophony of novel and increasingly interconnected ways. As we continue to do this, we will have to fundamentally alter the way we see each other. Connections will be much more about shared interests and shared experience as opposed to shared nationality.

Traditionally, there have been three interconnected but ultimately separate groups involved with the Irish diaspora:

the Irish at home, the Irish abroad and the Irish by ancestral links. There is an entire book to be written about the interplay of these in terms of past, present and future, but for now I will say that the dynamic is in an unprecedented state of flux. The Irish abroad are making their presence felt with the Global Irish Economic Forum – the Irish networks internationally and the Ancestral Irish are starting to engage with the state like never before. It will, therefore, be up to our generation to define these new relationships; how we see ourselves here in Ireland and how we see each other internationally is yet to be decided upon.

Already there are changes. At the onset of the recession there was genuine shame about being forced to leave the country, but more often there is an appreciation for the opportunities this affords. Particularly among young people, the conversations from friends abroad have led them to question the validity of staying in Ireland at all. Recent works by Irish-based artists have explored the ambivalence of the young Irish at home: the feeling of pride in staying and the feeling of shame from potentially being left behind. Nationalism is at its best when felt with pride, but in the future, the source of that pride is going to have to be renegotiated. In the great Junior Certificate stalwart, John B. Keane's *The Field*, we see how the source of pride rested in land, in a place. Since being 'Irish' in the future will be less and less dependent on a person actually living in Ireland, we will have to look at the source of our pride again. Being proud to be Irish will have to become more about great ideals than a shared homeland; it must become more about shared experience. When I was doing my J1 summer in California, our neighbour Bob brought out a tapestry that his grandmother had made for his grandfather. It was a perfect map of Ireland, with the four provinces delineated and each emblazoned with their respective crests in silver thread. The pride in Ireland his grandparents felt and the pride he felt via them was something to behold. That sense of pride is something we as a nation should look to exemplify in the future and learn to respect more.

How we look at the ancestral Irish is a key issue for our generation. For too long the 'born Irish' have looked upon their cousins (particularly the Irish-Americans) with a sort of 'paternal condescendence', implying that these people have a simplistic and over-generalised view of Ireland and that they love all parts of the island equally, with little appreciation for the nuances of intercommunity rivalry or differences. As the ancestral Irish become more diverse, more numerous and more connected to Ireland, we will need to appreciate them on a much more equal standing. The greater diversity for places of migration will also lead to a much wider proliferation of ancestral Irish. Irish-America is becoming more disconnected as they move into third, fourth and fifth generations. However, Irish-Canadian, Irish-Australian, Irish-German, etc. are all going to become major communities in themselves. We will have to learn to deal with people who are 'Irish' but may not speak English as a native tongue.

The major difference I envision as we move forward will be the move away from looking at the diaspora in terms of the state – Ireland – and more in terms of closer ties, such as the community a person grows up in (Dublin diaspora, Cork diaspora or Donegal diaspora) along with diasporas of interest (GAA diaspora, IT diaspora or *Father Ted* diaspora). The reimagining of the term 'diaspora' will blow apart its current limitations.

All too often have I met people who don't consider themselves a part of the diaspora because what they believe the term to mean is not what they believe themselves to be. This dissonance between semantics and reality is a great opportunity. It is our generation and it is our term. *We* will define its meaning in the future and no matter what we make it into, it looks set to be much more open, friendly and interconnected. It will be less reliant on nationality and more about the individual. It will be all of these things in a way no previous generation could possibly have imagined. Nowadays we can move abroad with very little difficulty and we can connect with friends and colleagues more quickly, more professionally and more easily than ever before. This truly is an exciting time.

Notes

1 Miller, K. (1985), *Emigrants and Exiles, Ireland and the Irish Exodus to North America.* New York: Oxford University Press.

2 <http://www.ey.com/GL/en/Issues/Driving-growth/Globalization---Looking-beyond-the-obvious---2012-Index>.

Chapter 9

Robert Nielsen

Economics I

With ambitions to be an economist or writer, Robert was winner of the Business and Economics category of The Undergraduate Awards in 2012 and he has just graduated from University College Dublin with a BA in Economics and Politics. One of the striking things about Robert is that he wanted to study these topics in order to help solve the economic crisis. He blogs regularly and he has featured on student and other radio programmes (though he insists he is quiet and thoughtful). Robert exemplifies the mindset of his generation, which believes that past mistakes borne out of greed should never be repeated. His vision is for a more fair and just society.

As an economics student, the first question I am invariably asked is: 'How are we going to fix the economy?' Or, to put it another way: 'How do we end the recession?' It hardly needs to be said, but Ireland's economy is in a deep crisis: unemployment is high, growth is stagnant and emigration is rising. As well as this, we are being crushed by the weight of the bank debt.

It may be surprising that I am somewhat optimistic. This crisis shows that the old ways have failed and that we are on the cusp of new discoveries. The old model is discredited and fit to be dumped and a new one must be built in its place. It is exciting to know that the world is about to change and there are vast possibilities for breakthroughs. We are passing through a unique window of opportunity to change the view of the world. We must not let this chance slip.

We have been here before. This is not uncharted territory where we must blindly wander. We can learn from the examples of the past. This is the worst crisis since the Great Depression – so surely we should examine the solution to *that* depression to find clues to solving our current one. What worked then may well work now. How was the last spell of mass unemployment solved?

The solution to the Great Depression of the 1930s was provided by John Maynard Keynes, the legendary British economist. He argued not for austerity but for stimulus. For Keynes, the boom, not the bust, was the time for cutbacks. When the economy is weak and private businesses are not spending, then the government must take its place by spending and investing when private business will not.

How does this work? How does stimulus help the economy and austerity harm it? In any given society, people and their actions are interconnected. When you spend money you provide income for a business that can then spend that money on wages, which is then spent by employees and becomes income for another business and so on. It is because of this interconnectedness that massive job losses in the construction industry brought down the rest of the economy. If fewer people buy houses then there is less demand for houses, so construction firms don't hire people, so more people become unemployed, so they spend less money generally, which affects the rest of the economy. This is how the economy so quickly went into a downward spiral. It becomes a vicious cycle.

How do we break the cycle? Surely if less spending caused it, then more spending is the solution. But who will do the spending? The economy will only recover if we all spend; if only one person does it, this person uses up their money and the economy doesn't recover. So everyone wants the other person to spend and so, as a result, no one spends and the economy stagnates. This is where the government steps in, not because it is the best option, but because it is the only option. The government is so large that when it spends money, this has a large effect on the economy.

But didn't spending get us into this mess? How can that be the solution? Economics is not a morality play: we do not prosper through being virtuous and pure, nor does fate punish the wicked. There is no cure-all solution, what is today's problem may be tomorrow's solution. We must spend when times are tough, not because that is our preferred option but because it is the only proven way of getting out of a recession. If consumers won't spend, then governments must. It is during the good times that we can afford to be thrifty and save money; in tough times we have to dip into our savings.

A good analogy is a car going up a hill. It must hit the accelerator going up the hill and tap the breaks on the way down. Regardless of how much fuel is in the tank, this is the option the car must take. It is not a question of preferred course or ideology but rather what must be done. The government's actions have been to tap the break going up the hill and, as a result, the car has stalled.

You cannot balance the budget with 14 per cent unemployment. Attempts to do so are like trying to push a boulder up a hill: every time you push it up a bit, it rolls back down again. When such a large proportion of the workforce is being wasted in forced idleness, the economy will not function properly. While unemployed, people are costing the state in welfare payments and their lack of employment also means that they aren't paying income tax. By enabling these people to work, not only does the government help to restore the individual's sense of dignity and self-worth but it also significantly boosts the economy and moves us all towards a balanced budget.

But where will the money come from? While it may seem unrealistic to think we can spend our way to prosperity, we can't afford not to. The record shows that if we don't spend, then the economy will stagnate. The short answer is: we must borrow the money. Now, getting ourselves into debt hardly seems like a solution but these are drastic times and the current financial situation makes it a necessity. Imagine if a flood damaged your house. The solution isn't to balance your budget but to fix your house, even if that means going into debt. You can then repay the loan during the good times.

But what about the markets? Well, the reason we are effectively cut off from the markets is because of the bank bailout. The market interest rate is based on the likelihood we will repay our debts. If you look at a chart of Irish bond rates, they jumped every time we poured money into the banks (the nationalisation of Anglo and the first and second recapitalisations of the banks coinciding with the largest jumps). The obvious solution is to stop wasting money on the banks. If €30 billion could be found for Anglo Irish Bank in exchange for no return or benefit and that again for the other banks, is it too much to ask for a small portion of that be spent on creating jobs?

We cannot carry the bank debt. I don't mean that it would be immoral to do so (although it would) or that it would be inefficient to do so (although it would) or that it goes against all major principles – politically, economically and socially (although it does). I mean it is a burden we *cannot* carry. Before we can make a strong recovery, we must ditch the debt. I do not say this out of ideology or lack of awareness of the consequences, but out of pragmatism. The ideas listed here may be called Keynesian but, in truth, they are the ideas of practicality. We must do them because they are the only option that will work. We cannot continue to flounder aimlessly without any direction. The solution I offer is not easy or simple but it provides a route to recover the state of the economy.

The world of economics is in a flux. The old walls are being broken down and even the very foundations are being questioned. Do the laws of supply and demand (those eternal laws always referenced but never proven) really describe the economy? Are people always paid a fair wage? Does the free market always lead to the best outcome? Can competition create more problems instead of solving them? Does the government always harm the economy or is it necessary for the survival of capitalism? Is the economy a dog-eat-dog world or do we have a responsibility to help the less well off?

Should society be judged by the glamour of the successful or by how it treats the powerless? Is greed always good?

A new Ireland would put greater emphasis on equality. For too many, prosperity is a dream and not a reality. If you are born into a working-class family, you automatically have a lower chance of going to university, earning as much or even living as long as someone from an upper-class family. There is enough wealth in the country to provide a basic standard of living for everyone; no one should be suffering on an island of want in a sea of plenty. This isn't just because it is wrong that opportunities in life be reserved for a minority, but because equality makes us *all* better off. Studies have found that more equal societies have less crime and less violence. In more equal societies, people live longer, have better mental health, feel less stress, have more trust, foster more community spirit and even have children who do better in school. Inequality was a major cause of the recession, since it led to an unbalanced economy and left people unable to buy goods to keep the factories running.

We must eliminate the scourge of unemployment. People are defined by their work: it's who they are. Unemployed people can suffer enormously from a loss of identity. They feel they are a burden and they lose their sense of dignity and self-respect. The shock of losing a job can be so damaging that many unemployed people suffer damage to their mental health and wellbeing. It is an enormous waste to see people with skills who are willing to work left to rust in enforced idleness.

Society is based on the principle of democracy – an idea we all admire and support. We believe democracy is the best way to run the country and yet democracy seems to stop at the door of our workplaces. What we need is economic democracy, applied both at work and at home. We spend a considerable portion of our days (a considerable portion of our lives) at work – is it that unreasonable that we have some say in what we do? Imagine if company decisions were decided by voting. Would it be chaos? After all, that was what was expected for political democracy. However, by allowing more people to participate, the decisions taken are always seen as fairer. In feeling that our opinions matter, our morale rises. By taking ownership of an idea, we work harder to make it succeed. Undeniably, it takes longer to make these decisions, but once they are taken they have a better chance of working because the majority is behind them.

I am, in effect, describing a co-operative, a model economists are increasingly examining to see what benefits it holds. Psychologists have long expostulated that the level of control someone possesses over their life greatly affects their life satisfaction and mental health. By giving people control over their work, they have a greater sense of purpose and achievement. Economic studies have shown that ideas like profit-sharing and collective action are successful because people work harder if they feel they have a stake in the business.

By giving people a share in the success of a business you are giving them a reason to work hard to ensure the business is a success. One of the major problems of a modern business is that staff feel disconnected and undervalued by their employers. By including staff in the discussion of the running of the business, this disconnection can be combated and ameliorated. Inclusion in a larger dialogue boosts employees' morale and productivity, as well as making them more loyal to the business and less likely to quit.

All around the globe people are experimenting with bringing democracy into the workplace. Companies are being set up where the staff have a say and have control over their work. The most famous example is of the Mondragon Co-Operative in the Basque Country. It is a collection of 250 co-operatives employing over 85,000 people. All decisions are taken collectively; this makes it easier to reach a goal, since everyone is pulling in the same direction. The highest-paid employee is paid only six times the salary of the lowest, a strong contrast with today's overpaid CEOs and their golden parachutes.

What about *how* we work? Instead of ruthlessly competing with each other and trying to work longer and longer hours in the office, how about we get a bit of work–life balance? Like Keynesian economics, it's a counterintuitive approach. But the reality is, if we want a more productive and successful society, we might need to spend *fewer* hours at the office. After all, work is only a means to an end – namely money – which is only a means to happiness and satisfaction. It hardly makes sense to make ourselves miserable working in order to make ourselves happy.

For most of history, people had to work extremely hard in order to scrape a bare existence. We are one of the few generations who do not. With all of our advances in technology, things that once took a week to produce can now be produced in a day. So why is it that we spend more and more time at the office? Are we so busy chasing the dollar that we've forgotten that we are a society?

Economists traditionally assumed that more was better, that having three cakes was better than two; therefore, the richer we got the better we would be. However, alternative views are beginning to challenge this. Surveys show that although we are far richer than our parents and grandparents we are not that much happier. Simply having more money may not make us better off. What is particularly interesting is that psychologists have found that it is not absolute wealth that counts towards our wellbeing, but *relative* wealth. People would prefer to earn €50,000 and be the best-paid person in the company rather than earn €60,000 and be the lowest-paid. So getting richer doesn't count for much: it just raises our expectations.

The benefits of a shorter working week would be great. We would have more time for our friends, family and personal relationships. With more free time, we would be less stressed and more relaxed, leading not only to a longer life but a better quality of life. We would finally be able to enjoy the pleasures of life that we always wished for but never had time for. We could learn a musical instrument or a new language. We could write that book we've always dreamed about. If we could re-evaluate our notion of 'being productive', we would probably use fewer resources. Good timing, since our planet will quickly run out of resources if we keep doing things the old way.

We need to take a much longer view of things if we are to get out of the recession and back to prosperity. We need to adopt the proven methods of Keynesian stimulus to boost the economy. We need the government to take an active and encouraging role. But we need more than just short-term solutions. We need a fairer Ireland; not just a richer one. We have to defeat the blight of unemployment, which has crushed too many young people's hopes and dreams. We need to give people control over their lives and bring democracy to the workplace. We should work to live, not live to work.

And all of this starts with a vision. If we have a vision for the future we want, then we can go about setting the goals that will take us there. With a bright vision and clear goals, there is nothing beyond our reach.

Chapter 10

Oisin Hanrahan

Economics II

Oisin is co-founder of The Undergraduate Awards and he has been developing the programme in Ireland and internationally for the last three years. A WEF global shaper, he was a scholar at Trinity College Dublin (while also operating a property development company in Hungary) and obtained his MSc in Finance and Private Equity from the London School of Economics before commencing his MBA at Harvard. It was there that he met the co-founders of his latest venture, Handybook.com, which is growing from its headquarters in New York.

Trillions of dollars of value, millions of new jobs and hundreds of thousands of patents have been created by companies that didn't exist fifteen years ago. Companies like Google, Facebook and LinkedIn have created enormous value in a very short space of time; they've grown from tiny startup ideas to operations on a grand scale in about a decade. Google's revenues in 2012 were approximately the same as those of the Irish Revenue. Google was founded in 1998.

My vision for Ireland involves running our economy with the best practices and values of a startup. Let's create a 'startup economy'. Let's take the lessons from the brightest startups and apply them to how we think about running our economy every day.

What simple changes can we make to how we run the economy so that we can benefit from the lessons of the fastest-growing companies on the planet?

VISION AND VALUES

Firstly, we need to project a big vision – a really big one – and choose values that go alongside the vision. 'What's the Irish economy like?' Eh... we need an answer to that question. We need a great answer and we should all be on the same page as to what it is.

What's the elevator pitch for the Irish economy? Every startup that is doing well has an elevator pitch that explains its competitive advantage and its importance in the world, succinctly. The process of coming up with an elevator pitch is difficult and time-consuming. But the value derived from doing it is incredible. Without an elevator pitch you can't explain why the world should care about your business. It allows people to quickly categorise your organisation in their mind and place you next to other similar businesses and it makes you more memorable. It also serves the basic purpose of making it easy for people to understand what you do.

As an economy we should have a pitch. We should have a pitch that is crisp, clean, simple and really compelling. I remember taking a class on the Irish economy in college and I recall the bullet points used to explain why it was competitive: low taxation; an educated, English-speaking workforce; and light regulation. These things may be true but they are not compelling.

The good startups have simple elevator pitches; the great startups have elevator pitches that are memorable, compelling and inspiring. We need to spend some resources plotting a vision for the economy so that we can distil it into a five-second version and a sixty-second version, and we should educate every person representing Ireland everywhere so that they know what this vision is. It should be so compelling that it's viral. It should be a vision that we're proud of.

The beauty, importance and power of getting the elevator pitch right should not be underestimated, since it becomes a guiding light that people will use to make decisions.

Similarly, we should spend some time outlining what the values for our economy are. These values should be the beacon that guides us through making tough decisions. These values will represent what we believe the important issues are and how we weigh different interests when critical choices have to be made. I believe that some of the values for our economy should include openness to innovation and change, competitiveness and progress.

CULTURE IS KING

Secondly, at startups there is generally a 'Be Cool' rule, because everyone knows culture is king. It is a rule that is fairly self-explanatory. The logic is simple: running a startup is hard and it involves working long hours in a stressful environment in close contact with other people. You want to be sure that you're working with people who you want to spend time with.

Our civil service needs to have a culture of success, achievement and brilliance associated with it. My vision is that we have a workforce of civil servants who themselves are not only proud to represent Ireland but, in turn, as a nation, every individual is proud of our civil service.

Working in the civil service should be a badge of honour, accomplishment and excellence. The best, the brightest and the most motivated should want to be a part of that team.

This is a long-term change – a change that requires us to adjust our 'job for life' approach to the civil service. We need to acknowledge that even the best people are occasionally going to make bad hiring decisions and when they do, we should allow them to make firing decisions.

If we're going to have a civil service we're proud to call our own, then it needs to have a culture of brilliance and the ability to fire people that don't fit that culture.

NO EXCEPTIONS: EXCEPTIONS ARE THE END OF EVERYTHING

We can't have exceptions to our values. I recently met Matthew Prince, the founder of CloudFlare, who articulated this problem very well.

Matthew told me that he once hired an engineer who insisted on a particular type of stand-up desk, when everyone else in the office had a $40 Ikea desk. Because Matthew really wanted to hire this person, he allowed him to be special and to get a different desk. Obviously this resulted in desk-envy and subsequent staff wrangling for special treatment. Soon, the conversation in the office wasn't about how they were going to get CloudFlare to grow faster, it was about what was fair or unfair about desks. The sum of the story was: exceptions are bad.

How does this relate to the Irish economy? Special interest groups and favours.

If we assume that somewhere in our vision for Ireland we specify that innovation and progress are key parts of our values, then we must uphold them. We must uphold our values in the face of opposition from lobby groups, from special interest groups and from vested interests. And we should know going in that it will be hard, because there are times that we'll want the votes, or we think we'll need the support. But if we pick the right values to begin with, then it is worth standing by them. If we set our values right, we really can't go wrong.

In my opinion, one of the bravest things the government has done is to stand by its belief in competition and free markets when adjusting the taxi market. I think it has delivered a taxi experience far better for customers. I commend the government of the day on making that incredibly tough decision to focus on the value of the free market in the face of opposition.

By the way, Matthew fired the engineer. When it comes to the values we set for our economy we should stand by them and make no exceptions.

SALES OVER EVERYTHING ELSE

No startup ever died because it sold too much. No economy ever blew up because it exported too much.

I remember when I was a kid and there was an ad on TV for Smarties, with the tagline: 'Only Smarties have the answer'. Every person in Ireland should know and understand that 'only exports are the answer'. As a nation, we're not all going to understand economics in a deep and meaningful way.

We're not all going to understand interest rate parity, ISLM curves and the monetary policy trilemma. And that's fine. But we all should know and understand enough to be able to make positive individual decisions that have a multiplicative, positive impact on the collective.

Put simply: people should know and understand enough about the economy in order to help it.

In the same way that years ago we ran a Buy Irish campaign, we need to run an exports campaign. We should be educating people that the most helpful thing they can do is to focus on encouraging exports. We should be celebrating every entrepreneur that runs export-driven businesses. Ireland is full of entrepreneurs that serve the local economy – but why stop there? The real economic heroes are the people who run businesses in Ireland in order to sell goods and services to people all over the world.

In the same way that in a startup we celebrate the sales people/user acquisition team and the marketers that bring in business, as a nation we need to celebrate those who bring home the bacon. Our exporters are bringing home the bacon. We should let them know how grateful we are.

STAY HUMBLE – TEST, ITERATE, TEST, ITERATE

In a startup, we come to work every day knowing that we can't predict what effect our actions will have. As a result we try not to take big actions without testing them first. So we make little tests. Lots and lots of little tests. We test everything maniacally. It's called minimum viable product testing.

Why don't we take this approach with government? Instead of driving policy based on ego, local constituency impact and politics, why don't we drive policy based on the results from small tests? What if we insisted on it?

Imagine if, for all capital spending over a certain threshold, a small test was required to show the expected results first. The initial reaction is generally that this will slow people down and force them into a process that makes progress and adjustment more gradual. However, if you think it through, it actually allows you to make much bigger, bolder and braver decisions. Because the amounts required for testing are relatively small, you can test large, revolutionary changes on a small scale. And if these tests show positive results, our decision-makers have the confidence and ability to propose bold moves with back-up data.

The effect of test-driven decision-making is huge. It allows us to change the option plane from 'what is politically acceptable' to 'everything we can test'. This shift in thinking is enormous and gives our government and civil service the opportunity to unearth and identify creative and innovative ways to solve the problems we face every day.

DATA VISIBILITY vs DATA AVAILABILITY

There is a big difference between data being available to your team and data being very visibly displayed and communicated in a way that everyone on the team understands, appreciates and can act upon. Most successful startups have a dashboard in their office that shows in real time how the company is performing. It might give daily displays of revenue, new users acquired, site up-time or user engagement. Beside each number it probably shows a green or red percentage sign to show how it is performing relative to the previous day or week.

The effect of centralising the important metrics and displaying them in a very clear manner to the team makes it very easy for the individuals to know how the company is doing and whether things are getting better or worse.

My vision for the Irish economy includes an economic dashboard with around ten metrics that encompass how the economy is doing. These metrics are easily displayed and understood by the entire nation. The dashboard is published on a weekly basis in the national papers and there is a website that users can visit to quickly view the metrics and to understand how, as an economy, we are performing.

The difference between having the data available but buried in reports versus having it clearly displayed and carefully curated is huge. My vision for the Irish economy includes a consistent and clear communication format called The Economic Dashboard, which everyone understands.

The importance of the economy to the lives of every individual in the country is such that we should ensure that everyone understands whether increasing inflation is good or bad. And this can be accomplished so easily by using a traffic-light system of green, orange and red to display the change in metrics.

It's obviously important to pick all-encompassing metrics that represent the short- and long-term benefit. For example, during the Celtic Tiger boom it would not have been sufficient to measure growth/unemployment, it would have been just as important to pick another metric focused on housing construction/debt to ensure that we were measuring the right things. Also, we should acknowledge that we may not get the metrics right the first time – but we can change what we measure if we find we're focusing on the wrong metrics.

REWARD THE PEOPLE WHO DO MORE EVERY DAY

After the team's engagement, belief and buy-in on the vision, the next biggest impact on success in a startup is the motivation of people to try more things. And the biggest factor in motivating those people is to ensure that they are appropriately incentivised. In any startup, the team is firstly motivated by an ownership incentive in the form of stock options and secondly through bonus stock and occasionally cash payouts linked to performance. It is obviously important to link the performance incentives to metrics that the individuals are responsible for influencing.

Let's link payment to performance. Let's motivate and reward our public servants that have a positive impact on the economy. Let's link payment to targets. The entire staff of the Department of Finance should have targets for the year that encompass the impact they are responsible for having. Similarly the Department of Jobs, Enterprise and Innovation should have targets focusing on unemployment, training, etc. These targets should be simple and easy for everyone in the department to understand. The power of putting the metrics on the wall will be incredible.

VALUES

The best startups are as much about having values as adding value. It is through our values that we create a vision. A clear vision is what's needed to look towards the future. So what's our vision for Ireland's future? My vision is for an Ireland that is filled with people, companies and policy-makers who are open to new thinking. My vision is for an Ireland that embraces technology, change and progress. My vision is for an Ireland that is proud of what we build, what we create and how we treat every individual.

Chapter 11

Tom Dillon

Farming

Upon graduating from his studies in Economics at Trinity College Dublin, Tom began managing his family farm, a mixed suckler and tillage enterprise in Meath. He has worked on farms across the US and the UK to gain a wider understanding of the differences in methodology between Ireland and the rest of the world, and he believes education and innovation are imperative to changing the views and practices of farmers in Ireland. In this essay, he makes a case for more young entrepreneurs entering the farming industry.

While farming has certainly been modernised since the first seeds were thrown into the mud of The Fertile Crescent around 10,000 years ago, at its heart lies the same simple principle. A *sine qua non* to our development as a species, the evolution and adaptation of farming has been the driver behind other advancements in civilisation. Once again, global agriculture is faced with challenges to which it must adapt and evolve. Farmers can expect much upheaval between now and 2050.

Farming in Ireland appears in many positive news stories, since the agricultural sector has been identified as an economic anchor: an indigenous industry that remains resolute in the face of turmoil. In the national effort to export out of financial crisis, farmers and food producers have been extremely effective. The agri-food sector currently accounts for 10.8 per cent of Ireland's total exports, 25 per cent of all net foreign earnings and 8 per cent of total employment. In 2012, Ireland's Gross Agricultural Output was valued at €6.61 billion. Breaking this figure down, we see that beef production accounted for 39 per cent of national output or around 495,000 tonnes, of which approximately €1.9 billion was exported. Ireland also exported approximately €205 million of sheep meat and approximately €506 million of pork. Total milk output was estimated at 5,200 million litres. There are around 370,000 hectares under cropping, which includes cereals and vegetables; and farmers and producers received €1.2 billion in Common Agricultural Policy (CAP) funding.[1] The industry is clearly in a strong position, competing successfully on a global scale by selling into diverse markets that remain keen to buy our produce. Couple the industry's export growth with increasing world food prices and so far it is positive news indeed.

However, the Irish farming sector faces many challenges also, both economically and socially. Economically, the cost of food production is rising dramatically, outstripping any food price gain at farm-gate level. Energy prices are approaching unsustainable levels and the effects of climate change are forcing changes in farm management while also generating massive volatility in food markets. Socially, the mean age of rural Irish communities is increasing. Just 7 per cent of

Irish farmers are under the age of 35. This may be greater than the EU average of just 6 per cent (this figure drops to as low as 3 per cent in some member states)[2] but it is still not a position of strength. The issue is not a lack of interest in farming from young people exclusively, but rather a number of problems that plague most young entrepreneurs: access to land, low return on investment in the initial stages and issues with raising finance. Emigration to countries such as New Zealand, which offers its farming industry much easier investment access,[3] is not just draining the sector of talent but is adding to social issues such as isolation as rural communities decline.

Food commodity markets have recently entered a period of high volatility. It is not uncommon for Irish farm output to see over 100 per cent price increases one year and equivalent decreases the next.[4] This market volatility not only scares people, but also creates uncertainty over future price levels and this complicates long-term planning. This in turn leads to producers of commodities underinvesting in the physical assets that support growth.

Historically, European Union farm policy created stability in food prices to purposefully avoid these issues. But the decoupling of the CAP payments from output, which occurred in 2003, saw this mechanism for stability removed.

An examination of the causation of this volatility reveals some of the predominant issues that are facing both global and Irish agriculture. These issues include the economic growth of several large developing nations, the effect of this growth on the price of oil and on food demand, the increase in weather shocks related to climate change, the stagnation in cereal yields and the availability of water.

FOOD DEMAND AND THE PRICE OF OIL

The economic progress of large developing countries such as China and India has driven demand for oil while simultaneously driving world demand for food; especially a more protein-based Western diet that includes more meat and dairy products. Considering we export 85 per cent and 80 per cent respectively of all our meat and dairy production, this is a boon for Ireland, which is further enhanced by the fact that Ireland's agricultural output is based primarily around grass production and the transformation of this natural resource into proteins via ruminating animals. The efficacy of Ireland to grow grass has made us the most carbon-efficient milk producer in the EU and the fifth most carbon-efficient beef producer in the EU.[5] These statistics will also be shown to be the source of our greatest competitive advantage going forward.

Unfortunately, the economic success of the developing world that has driven the demand for food has also driven demand for energy. This in turn has raised the costs of food production greatly. Global food production is heavily dependent on fossil fuels. The 'Green Revolution' that agriculture experienced from the 1940s through to the 1970s relied heavily on cheap oil, to the point that intensive farming practices essentially implied the conversion of fossil fuels into food. Oil and gas are the major ingredients for the synthesis of chemical fertilisers and pesticides. They are also the primary energy source at all stages of food production: from planting, irrigation and harvesting, through to processing, distribution and packaging. Additionally, fossil fuels are essential in the construction of farm equipment and the infrastructure needed to facilitate the industry, e.g. machinery, processing factories, storage units, ships, transport vehicles and roads.

Independent economic think-and-do-tank, the New Economic Foundation (NEF) published a report hypothesising that we have already seen the end of cheaply available oil, which also means we will have seen the end of cheap food unless breakthroughs in biomass or in alternative energy are made rapidly. The International Energy Agency, the International Monetary Fund and the G7 have warned that high oil prices have likely been constraining global economic recovery. The NEF defines this as 'economic peak oil': 'the point at which the cost of incremental supply exceeds the price economies can pay without significantly disrupting economic activity at a given point in time.'[6]

It is evident that this hypothesis could prove correct. The hardships in the agri-industry and the haulage industry, as well as the struggles for commuters and homeowners trying to keep warm in this country are very visible. The economic crisis, therefore, is related to the cost and availability of transport fuels – fuels that account for 80 per cent of all oil usage.

Transport fuels link all elements of the food supply system. If every linkage costs more because of sustained high oil prices, all costs will increase, the economy will slow and inflation will rise. If a larger proportion of people's incomes are dedicated to food and energy costs, then it will be increasingly difficult to create growth in the wider economy. Such an eventuality will not be good news for Ireland, which (in spite of being a major beef and milk exporter) currently imports 60 per cent of all our food. Further rises in fuel costs would signal the end of what was an extended period of agricultural over-production worldwide and the beginning of long-term scarcity in food markets.

For the Irish food-producing sector, these potential increases in transportation costs may cause our primary exports to lose competitiveness. They would especially impact on the beef and sheep industry, where our meat exports are marketed as being low-cost. (This is an unusual marketing strategy for a country in the developed world, especially when we could and should be focusing on added value rather than undercutting our neighbours. Unfortunately it is a historical product of poor administration and a lack of confidence in our own abilities.)

Luckily for our food exporters, however, our largest market is also our nearest neighbour: 42.5 per cent of all our agri-food exports are to Britain. This is followed by France, which annually buys €645 million worth of our food exports.[7] This geographic proximity to our major markets should assist in mitigating any crisis in export transport costs while also adding a competitive advantage over the other major world beef exporters who sell into Europe, as they are located primarily in South America. The predominantly urban population of the UK generates a demand for beef which it cannot self-sustain. Although British beef farmers may not appreciate the march of beef across the Irish sea they will not be in a position to fulfil demand for beef in an affordable manner for the British consumer for quite some time. The high welfare standards of Irish livestock, matching those of our British counterparts, will also ensure that any substitution for another beef supplier is unlikely.

GLOBAL WEATHER SHOCKS

We can now assume that the outcome of anthropomorphic climate change is underway (despite a resistance to accept its findings in some quarters). We may not be able to easily identify the incremental rise in temperatures or the gradual increase in annual precipitation but we can easily identify the increased occurrence of weather shocks. Weather shocks – random weather-based disasters such as droughts, floods, storms and blizzards – result in supply shocks, since they directly impact on crop yields and animal vitality. Weather events affect prices, incomes and land use. Recent examples of weather-related supply shocks include the unprecedented heat wave in Ukraine, Kazakhstan and Russia in 2011, which caused global wheat supply to drop by 5 per cent; and the drought of 2012 in the US Midwest, which resulted in a corn price hike of over 50 per cent. At the time of writing, parts of Ireland are experiencing their second serious flood in nine months.

Although difficult to predict, the forecasted change in Ireland's climate is for a mean temperature increase, fewer frost days annually and an increase in rainfall amounts in the west of the country with a slight decrease in rainfall in the east. The Environmental Protection Agency is not expecting the need for any significant crop or management changes for Irish farms in its predication models.[8]

In fact, the Intergovernmental Panel on Climate Change predicts that Northern Europe may actually benefit. In Ireland the current data suggests that, outside of any repeated catastrophic weather shocks, climate change will result in either an increase in productive farm output or, if yield increases are not expected for certain crops, significant reductions in nitrogen necessity.[9] These reductions will be realised through the increased CO_2 levels in the atmosphere improving nitrogen usage and also the reduced rainfall in the months from May to September reducing the effects of leaching. Although temperatures may rise in the summer months and we may experience milder winters, there is no reason to believe that this will lead to any major issues with the availability of water for irrigation or an improvement in the environmental conditions for weed growth or pest proliferation. Ireland's reputation as a reliable source of food will be maintained and, at the same time, the possibility of greater yield exists. Although it is not possible to quantify this benefit financially at present, it creates the opportunity for Ireland to enter more markets in the future and also to supply them with a sustainable production growth.

It is a very different story in other regions worldwide. Southern Europe, Africa, South Asia, Latin America, Australia and New Zealand are all projected to undergo significant yield reductions primarily as a result of drought.[10] These droughts will certainly be exacerbated by the effects of increased water demand due to urban expansion and population growth. Serious farm management changes will be necessary to reduce the impact of yield reduction in the most affected regions. It may even result in the abandonment of what were once staple food sources and the rapid increase in the incidence of hunger.

SUSTAINABILITY

With worldwide demand for beef and dairy to increase by 110 per cent and 85 per cent respectively by 2050,[11] and with climate change to have negative yield impacts across much of the planet, the sustainability of current agricultural practices needs to be examined.

Globally, agriculture accounts for a quarter of all greenhouse gas emissions but only 4 per cent of the world's GDP, meaning agriculture is unfairly categorised as extremely emissions-intensive. This figure is biased: of all the types of human activity on the planet, food growth and its marketing are surely among the most vital. Food prices have remained stubbornly inelastic in the developed world because of recent overproduction and the dominance of subsistence agriculture in the developing world; this means that the sector's importance is poorly reflected in GDP.

In Ireland, a similar statistical anomaly exists. As Teagasc recently noted in their report on Irish greenhouse gases and climate change, Irish agriculture accounts for 30 per cent of all greenhouse gas emissions nationally. This compares badly against the EU average of 9 per cent; however, this is an unfair ratio, since these statistics are calculated territorially and not on productive output. Irish agriculture is producing enough food at present to sustain 34 million people. When productivity is taken into account, Ireland has one of the lowest co2/kg ratios on beef and dairy output internationally.[12]

As oil and farm input prices continue to increase, the market is dictating that further efficiencies will have to be achieved. Ireland has been foremost in doing this; farmers have helped reduce their environmental impact by increasing output significantly while leaving their carbon footprint static. Efficiencies in nitrogen use, the use of organic manure, improvements in grass management and developments in the Economic Breeding Index of livestock have all helped to improve output without a correlative increase in co2 emissions.

Consideration should also be paid to Ireland's predominantly permanent pastureland usage and the carbon sink nature of grassland. With grass-based systems dominating livestock production in Ireland, inclusion of grassland sequestration should actually put Ireland at a significant advantage in comparison to other major European producers.

FROM FARM TO FORK

There has been much media comment of late about the massive length of the current food supply chain and the disconnection between consumers and the origins of the food they eat. I don't believe this situation will last past the short-term. Food prices will continue their volatile climb affecting the poorest first and then eventually the developed world. This will generate increased interest in what exactly consumers are paying for, especially if consumers are forced to start substituting certain staples for cheaper alternatives in their weekly food shopping. The idea of certain basic food items becoming unaffordable for Irish consumers may seem absurd now, but the current situation of taking food supply for granted is unsustainable. The over-production that epitomised the 'Green Revolution' has ended in stagnation in yield growth. An expectation of continued food output growth to meet population growth is no longer possible. The reliance of world food production on fossil fuels is increasingly unsustainable as is the expectation that weather systems will continue behaving as they have been in the recent past.

Perhaps by 'the luck of the Irish', our agricultural industry will be well placed globally in terms of the sustainability and efficiency of our production methods, our geographic location, the predicted effects of climate change and our proximity to our primary markets. This will be of great financial benefit to the economy and of great benefit to the nation's food security in the future. Not everywhere can be so lucky, and it's an unfortunate truism of the modern food markets that one farmer's misfortune becomes another farmer's gain. The issues facing global food production are complex and any solution will have to be equally complex.

Ireland's future role as a food producer may well encompass crop changes. This will not be because of climate change affecting the Northern European region (as noted, it largely won't) or because of sustainability reasons, but as a larger worldwide response to climate change. Ireland's ample water resources could lead to an increased production of water-thirsty crops such as corn here. This will reduce the need to grow these crops in the more arid regions of the planet where sustainability issues are more likely to arise. In turn, the hotter and drier regions can concentrate on growing more drought-resistant cereals.

Irish agriculture has always been one of the quickest to adapt and adopt new methods and techniques; this cultural openness to change has made our industry a world leader. World agriculture will be very different by 2050 but Ireland can continue to exploit its natural and cultural advantages for the nation's benefit. It may even create enough interest to attract a few under-35s.

Notes

1 <http://www.teagasc.ie/agrifood>.

2 Bogue, P. (2012), *Potential of Farm Partnerships: to Facilitate Entry into and Establishment in Farming*. Ireland: National Rural Network, p.1.

3 Ibid. 8.

4 Newenham, P. (2012), 'Price of potatoes soared 177% over 12 months', *The Irish Times*, 17 October.

5 Teagasc (2011), 'Ireland Best in EU for Carbon Footprint of Milk, Pork, and Poultry Meat', <http://www.teagasc.ie/news/2011/201102-15.asp>.

6 Johnson, V., Simms, A., Skrebowski, C. and Greenham, T. (2012), *The Economics of Oil Dependence: A Glass Ceiling to Recovery*. New Economic Foundation.

7 Bord Bia (2012), '*Factsheet on Irish Agriculture and the Food and Drink Sector*', <http://www.bordbia.ie/industryinfo/agri/pages/default.aspx>.

8 Sweeney, J. *et al.* (2008), 'Climate Change – Refining the Impacts for Ireland'. EPA: STRIVE Report.

9 Contribution of Working Group II to the Fourth Assessment Report of the Intergovernmental Panel on Climate Change [M.L. Parry *et al.* (eds.), *Climate Change 2007: Impacts, Adaptation and Vulnerability*. Cambridge University Press: IPCC 2007].

10 Ibid.

11 O'Mara, F. (2011), 'The Significance of Livestock as a Contributor to Global Greenhouse Gas Emissions Today and in the Near Future', *Animal Feed Science and Technology*, 166/23, 7.

12 Schulte, R.P.O. and Lanigan, G. (2011), 'Teagasc Working Group on Greenhouse Gas Emissions' (eds.), *Irish Agriculture, Greenhouse Gas Emissions and Climate Change: Opportunities, Obstacles and Proposed Solutions*. Teagasc.

Chapter 12

Julie Clarke

Infrastructure

Julie was a winner in the inaugural Engineering category of The Undergraduate Awards in 2009, and her continued research was subsequently recognised by the 2011 European Science, Engineering and Technology Awards. She is currently undertaking a PhD in University College Dublin, for which she was awarded a scholarship by the Irish Research Council. Her research focuses on Risk Assessment for Masonry Buildings Subjected to Tunnel Induced Ground Movements. She has been published widely on this topic and in 2012 she presented her research at the prestigious Falling Walls Lab in Berlin.

In order to provide a desirable quality of life for present and future generations, civil infrastructure is crucial. We rely on high-class road and railway systems to ensure safe transportation; dependable water treatment plants and water mains facilities to deliver a constant supply of clean drinking water; and an enduring energy supply to meet our heating and lighting needs. Day-to-day, it is easy to take these items for granted. Sometimes it is only the occurrence of a natural or manmade disaster that heightens our awareness of our dependency on such infrastructural systems.

Escalating global issues have amplified the need for robust infrastructural systems worldwide. In the past fifty years alone, we have seen the percentage of inhabitants living in urban areas almost double, and is it predicted that by 2050 two-thirds of the world's population will be living in cities.[1] We have also witnessed a significant rise in sea levels in the past decade, which has led to increased instances of flash flooding. Moreover, with the world's supply of fossil fuels rapidly depleting, the search for sustainable resources has become a vital task.

Ireland, however, is facing an age-old problem, one that many countries around the world are also facing at the moment. Our country's civil infrastructure is presently insufficient – a problem that will undoubtedly escalate if we do not take action now. A crucial element to the development of such systems is capital. But, quite frankly, there is no money. Since infrastructural investment today will not only avoid many long-term problems but will also facilitate economic recovery, we are faced with a predicament.

What are we to do? In a way, both sides need to give in a little. Yes, I believe our government should make infrastructure a priority – there is too much evidence to suggest that not doing so will be detrimental. However, civil engineers such as myself need to develop and propose low-cost solutions. This may require some creativity, and there's no harm in that.

Irish public capital spending on infrastructure-related civil projects has diminished significantly for reasons we are all too aware of. Government budgets pertaining to transport and water system infrastructure have been cut by almost half for the forthcoming three years, compared to preceding years. Furthermore, the majority of what funding remains is geared towards the maintenance of existing infrastructure rather than investment in new projects, despite the fact that our existing transport and water systems are presently considered to be inadequate.[2] In addition, several long-awaited, much needed infrastructural projects have been postponed for indefinite periods.

In addition to providing a means for meeting the needs of society, infrastructural investments can have significant positive effects on the economy, both immediately (by acting as a stimulus tool for job creation) as well as in the long-term (by strengthening the economy). For Ireland in particular, immediate investment would create much needed employment opportunities. In fact, estimates show that the employment intensity of projects ranges from eight to thirteen jobs for every €1 million invested.[3] In the long-term, infrastructural investment would serve to increase the competitiveness of the Irish economy. In fact, failure to invest in infrastructure has been known to result in significant adverse effects on a country's economy. Remarkably, the US predicted that a failure to maintain their surface transportation services would result in the suppression of their GDP by $897 billion by the year 2020.[4]

In the past five years, we have seen Ireland's construction sector contract by almost 80 per cent[5] and at present this sector comprises just 6 per cent of our total GNP.[6] Large sections of the trained engineering work force have either left the profession or left the country and unemployment rates, particularly amongst construction workers, are at an all-time high. Rebuilding Ireland's construction sector to the standard 12 per cent of GNP for growing western economies should be critical to our economic recovery plan.

The European Commission has highlighted the importance of maintained investment in infrastructure in order to sustain economic activity and support a return to growth.[7] What's more, despite the fact that significant investments were made during the Celtic Tiger era, Ireland is increasingly falling behind its European counterparts in terms of infrastructure provision: recently Ireland was ranked in twenty-fifth place out of twenty-eight countries according to quality of infrastructure.[8] Therefore, we must continue to maintain our infrastructure in order to remain a competitive economy. One does not need to look beyond the period following Ireland's decade of under-investment in the 1980s to understand the negative ramifications in terms of economic growth.

Owing to the present fiscal constraints, however, the development of efficient and cost-effective infrastructural solutions is critical. Specifically, it is crucial for civil engineers to provide cost-saving solutions to fulfil our current infrastructural deficit, and to provide a means for maintaining our overall infrastructural system in the future in order to satisfy the needs of succeeding generations. Moreover, it is our obligation to encourage and persuade governmental bodies, authorities and private developers to pursue infrastructural development and to convey the alternative grave consequences for our society. As such, my current doctoral research at University College Dublin's School of Civil, Structural and Environmental Engineering tackles the issue of the world's rising urban populations.

The problem of urban crowding is clearly evident in Tokyo city, where over 34 million people currently live. There are an additional twenty-one megacities worldwide that consist of a population of over 10 million. These extreme population shifts to urban areas, occurring in the last fifty years, result in increased pressure on transport services, water and waste systems, as well as utilities. Furthermore, they raise fundamental questions as to how such densely populated cities can deliver clean water to their millions of inhabitants or provide efficient transport where ground space is rapidly diminishing.

The answer to these questions lies in underground construction, particularly tunnelling. Tunnels provide the much-needed additional space in urban environments and allow for transportation of people, water, waste and utilities. Ultimately, tunnels enable these cities to remain liveable under increased population growth and denser urbanisation. These structures provide an opportunity for sustainable development in expanding cities by providing the necessary space for the passage of infrastructural systems.

Most notably, urban tunnelling enables us to move away from car-based city living. Astonishingly, in the UK it is estimated that traffic congestion results in annual economic losses in excess of £20 billion.[9] In addition, tunnels provide the opportunity for increased surface space, providing a solution to the worldwide request for more urban space generally. By locating infrastructure, services and utilities below ground level, valuable surface spaces may be reserved for public use. This allows for the protection of natural resources, such as land, water and bio-diverse environments, as well as reducing air pollution by removing surface traffic. (Although of little relevance to Ireland, tunnels are particularly suited for earthquake-prone cities, since these structures are resilient to earthquakes, enabling modern cities to cope with these increasing populations while simultaneously withstanding the effects of natural disasters.)

Unfortunately, lawsuits are a common feature of tunnel projects. This is due to the fact that tunnel excavation results in ground movements that may damage overlying buildings unless the appropriate building protection measures are implemented. These movements are caused by the subsoil stress release that occurs as the tunnel is excavated which then propagates to ground level in the form of surface settlements. This can subsequently cause the deformation of overlying buildings and may result in unacceptable damage levels. Although significant advances have occurred in tunnelling technologies in recent decades, the ability to reliably prevent building damage has not kept pace.

In the past decade alone, we have witnessed several catastrophes. In 2009 the collapse of Cologne's archive building, which contained important documents dating back as far as 922 AD, is believed to have been related to the construction of the adjacent Cologne Stadtbahn subway line. In addition, the construction of the Barcelona Metro extension in 2005 caused a major collapse, resulting in two apartment blocks being totally wiped out and more than fifty families left homeless following the incident.

However, it is not just catastrophic events such as these that we wish to avoid but also the prevalence of widespread, low-lying building damage that remains a common trait of tunnel projects. A recent example of this happened in Ireland during the construction of the Dublin Port Tunnel, which resulted in 334 uncontested property damage related claims,[10] adding significant cost to the project as a result of pay-outs. In order to prevent financial losses for future projects and to maintain support for tunnels, improvement of this situation is critical.

Since tunnel projects in cities are generally located beneath dense urban environments (in order to relieve congested areas), the settlement effects of tunnel excavation may impact on hundreds, if not thousands, of buildings. For most European cities, the bulk of their built heritage is formed of brickwork or stone buildings (i.e. masonry material). If we take Dublin city as an example, 91 per cent of buildings in the central Grafton Street area are comprised of either brickwork or stone or a combination of both.[11] Masonry buildings are particularly vulnerable to the effects of ground movements because of the inability of this building material to accommodate tensile strain, a phenomenon that occurs as these buildings undergo deformation due to the imposed ground surface settlements. For many of Europe's older buildings, these may be architecturally and/or culturally significant. For example, the aforementioned region in Dublin city consists of almost half of buildings classified as protected structures. If we consider that roughly 5 per cent of Europe's GNP is generated through tourism, the protection of these buildings during infrastructure development is imperative in terms of economics – not to mention culture.

In order to prevent tunnel-induced building damage, projects generally conduct a risk assessment of the overlying built environment prior to tunnel excavation. At present, risk assessments are commonly conducted as follows: (1) all buildings along the proposed tunnel route are surveyed prior to tunnel excavation in order to detect any existing defects and to identify those in poor structural condition; (2) a damage prediction is subsequently conducted for all buildings, based upon the proposed tunnel geometry, relative position of the building, building properties, as well as other considerations; (3) for buildings where unacceptable damage is anticipated, detailed assessments involving computer modelling are conducted and protection measures are specified and implemented prior to excavation.

When predicting building damage, there is frequently a large degree of uncertainty caused by the inherent variability in subsoil conditions and the frequent absence of building information. Notably, subsoil conditions influence the shape and magnitude of generated surface settlements and building properties may influence the overall response to these ground movements. Particularly for older buildings, there is a large degree of uncertainty present when making predictions. This is because building drawings are rarely available: consequently, structural layouts, building foundations type and material properties are difficult to determine because of the limitations of manual building inspections. In addition, these older buildings are often in a poor state of repair, possibly because of neglect or perhaps because of previous building movements (which are extremely difficult to discern for the majority of buildings). In addition, determination of the extent to which these movements have influenced the overall integrity of the building is near impossible.

In my view, the continued inability to accurately predict and thus prevent the occurrence of building damage during tunnel projects is due to the over-idealising of structures and the failure to quantify the inherent uncertainty. As such,

I am proposing a novel approach to the risk assessment of masonry buildings for tunnel-induced ground movements. This will adapt a technique that is currently used for the risk assessment of buildings to earthquakes.

The method that will be adapted accounts for the uncertainty in predicting the expected earthquake magnitude experienced by the building and also considers the uncertainty in determining building properties that may influence the building's response to the earthquake shaking. In order to do so, the method employs probabilistic techniques, enabling the uncertainties to be understood and quantified in mathematical terms. Although the traditional context of this technique is for life safety determination of buildings, this adaptation will enable a similar means for accounting for the uncertainty present when predicting the less extreme case of building damage due to tunnelling.

As such, my research is focused upon the prevention of low-level widespread building damage, as was the case following the construction of the Dublin Port Tunnel. My work is, in part, inspired by the well-known tunnel engineer Alan Muir Wood, who stated that 'uncertainty is a feature that is unavoidable in tunnelling... but it can be understood and controlled so that it does not cause damaging risk'. In addition, my work is inspired by the fact that earthquake engineers have demonstrated a proven ability to provide reliable building risk assessment techniques. At the outcome of my research, I aim to determine the probability of an individual building exceeding a specific damage level and, for large groups of structures, to determine overall likely building damage to be encountered because of the construction of an adjacent tunnel.

Ultimately, I hope to be able to reduce the number of damage-related lawsuits following tunnel construction and to provide a means for protecting our built heritage as we develop our infrastructure to meet the needs of expanding urban populations. Civil engineers must provide cost-saving solutions to remedy the discrepancies of Ireland's existing infrastructure. As a society, we can acknowledge the challenges we are likely to face in the future and provide action now to sustain a desirable quality of living for years to come.

Notes

1 Oliver, A. (2010), *The New Urban Dream.* New Civil Engineer.

2 Engineers Ireland (2012), *The State of Ireland 2012, A Review of Infrastructure in Ireland.*

3 Construction Industry Council (2009), *Submission to the Government by Construction Industry Council, Jobs and Infrastructure: A Plan for National Recovery.*

4 American Society of Civil Engineers (2011), *Failure to Act: The Economic Impact of Current Investment Trends in Surface Transportation Infrastructure.* Economic Development Research Group.

5 Construction Industry Council (2012), *Building Our Future Together.*

6 Society of Chartered Surveyors Ireland (2012), *The Irish Construction Industry in 2012.* DKM Economic Consultants.

7 Commission of The European Communities (2009), *Communication from the Commission to the European Parliament, the Council, the Economic and Social Committee and the Committee of the Regions. Mobilising Private and Public Investment for Recovery and Long Term Structural Change: Developing Public Private Partnerships.* Brussels.

8 IBEC/KPMG (2011), *What Next for Infrastructure? Infrastructure Insights for Ireland.*

9 Hall, P. and Pfeiffer, U. (2000), *Urban Future 21: A Global Agenda for Twenty-First Century Cities.* Spoon Press.

10 Brennan, M. (2012), 'Many Still Awaiting Tunnel Claim Payouts', *Irish Independent.*

11 Clarke, J. A. and Laefer, D. F. (2012), *Generation of a Building Typology for Risk Assessment Due to Urban Tunnelling.* Proceedings of the 2012 Joint Symposium on Bridge, Infrastructure and Concrete Research in Ireland. Dublin, pp.487–92.

Chapter 13

Sasha de Marigny

Multiculturalism

Sasha spent two years establishing The Undergraduate Awards internationally. She started as project co-ordinator and then moved into a communications role. She is now a senior associate at Ernst & Young, managing PR for the Entrepreneur of the Year awards. She was one of the founding members of the Dublin Sandbox hub, she is a WEF global shaper – and she is PR officer for both organisations. She also has a talent for acting and co-founded the Talentwest Performing Arts School in Athenry. A Galway girl, she originally came to Ireland via South Africa; and her essay focuses on the many so-called 'non-nationals' who now call this country home.

Home. It's a strange concept. One that interchanges throughout a lifetime, can exist in multiples and, for some, doesn't exist at all.

I came to live in Ireland from South Africa six years ago. While the decision to emigrate was a pretty impulsive one, I didn't just spin a globe and select the first place my finger landed on. I was an Irish citizen prior to even receiving my

South African identity book. My mother is a born and bred Galway girl: both her mother and father, and generation-upon-generation before them, hailed from the west of Ireland. My father, on the other hand, is French-Mauritian. Family occasions were a strange mishmash of Irish accents and creole-French with a peppering of 'S'effrican' – an environment, as you can imagine, that wasn't exactly conducive to the development of a staunch national pride. In short, I was born a mutt: a mutt with a Russian name, a French surname and an Irish passport, living in an African city that was colonised by the British in a country with eleven official languages.

But I'm not the only one. South Africa is the epitome of a cultural melting pot; a land of approximately 52 million mutts. A 'rainbow nation' with a population that is 79 per cent black African, 9 per cent coloured (a term with very specific meaning in South African demographics), 9 per cent white and 3 per cent Asian. I grew up, blonde hair and pale skin, in KwaZulu Natal, the Kingdom of the Zulu – a humid, sub-tropical city that is known to have the largest Indian population outside of India. I was raised in one of the 10 per cent of South African households that speak English as a first language, my mother tongue ranking fourth behind Zulu, Xhosa and Afrikaans. Despite all of this, I was completely at home, nestled amidst this vastly heterogeneous society and happily maintaining my mongrel-esque identity of being a half-Irish-half-French first generation South African.

Fresh off the boat in Galway city, people and places became 'grand' rather swiftly. Soon the word 'like' started to pop up at the end of my sentences. It took a while, but I finally realised that the phrase 'yer man' wasn't in reference to my boyfriend.

To my understanding, these were the first tell-tale signs that I was starting to settle into my life in Ireland and leave my African ways behind. I was becoming one of you. Well, at least I thought so.

I am always slightly taken aback when people (taxi drivers, mainly) say to me, 'Well, *you're* definitely not one of us' or 'That's a funny accent' or worse, 'How are things Down Under?' (that's an Australian reference, by the way). For some bizarre reason, after being here for six years, I kind of felt that I had blended in by now. I had my nana, great-aunts and great-uncles and over a hundred cousins scattered across the west coast. Strangers would tell me stories about my great-grandmother and her shop on Dominick Street, relaying anecdotes about how she'd sold them single cigarettes when they were teenagers. This had never happened to me as a child in South Africa, so how could people refer to me as a 'foreigner' here?

On the South African front, it became a similar story. I was chastised by my friends for speaking weirdly. I was slapped on the wrist for how frequently I punctuated my speech with f-bombs. And I slowly but surely started to feel a disconnect with the South African bourgeois bubble I had once found so cosy – a feeling that was cemented when, out of the blue, on my last trip to South Africa I referred to departing for Dublin as 'going home'. So where did this leave me? If I no longer felt at home in South Africa, but was considered a foreigner in the place that I had begun to call home, did that render me homeless?

Over the past number of years, Ireland has evolved into a flourishing ecosystem for multinational megaliths such as Google, Facebook, Twitter and LinkedIn. The demand for a cross-section of native speakers to manage European and Middle Eastern clients has never been greater and, since Ireland is perfectly situated as a stepping stone between continental Europe, Asia and the US, it has truly given a hundred thousand welcomes to America's tech titans and the international workforces they bring along with them.

While large-scale emigration (particularly of young, talented Irish graduates) continuously dominates headlines, little attention is paid to the talent that is coming to this island from foreign shores. Renowned for hiring the very best and brightest, Google attracts employees to join their workforce from across the globe. Having worked in a startup based in Google for the past two years, I swiftly felt right at home within the diversity of the workforce, revelling in the opportunity to eavesdrop on conversations held in over seven different languages on a daily basis and witnessing first-hand the organic formation of multiple strong communities within the diverse workforce. No matter what part of the world you've come from to be in Dublin, your 'Googler' status instantly provides access to a vibrant, like-minded community of well-travelled, educated and ambitious individuals. Within the enclaves of this posse (and the flurry of social events that are part and parcel of being a Googler), immigrants are granted the ability to dodge the usual isolation and loneliness, surrounding themselves in a cushioning structure that serves as somewhat of a home away from home.

However, this magnetic desire to surround oneself with a semblance of 'home' and community, extends beyond the ivory tower of Google and its multinational compadres. Excluding the hermits among us, we're community creators and subscribers by nature, a trait even further bolstered by the digital age in which we live. Communities give one a sense of belonging, a sense of security and, most important, a sense of feeling a little more at home. Snapshots of small immigrant communities are evident across the island, from the early-stage Chinatown sprouting up on Dublin's Parnell Street to the 20,000-strong community of Brazilians inhabiting 'Little Brazil' in the relatively small town of Gort, Co. Galway. The 2012 census revealed that people born outside of Ireland make up more than two-thirds of the population around O'Connell Street, further demonstrating that in an age governed by mobility and connectivity, a time where 'home' is becoming a more transient concept, the need to roll with your homies burns ever strong. Now, more than ever, our attachments to places are beginning to loosen as our ties to people continue to strengthen.

!KE E: |XARRA ||KE – UNITY IN DIVERSITY

So what does this mean for an Ireland that is increasingly becoming 'home' to a myriad of communities from around the world? By its very location, Ireland has the potential to lead the way in demonstrating how an effective multicultural society can function – and thrive – by harnessing the unique offering each community has to provide to the nation, going beyond mere tolerance to cultivating a genuine understanding, curiosity and acceptance of living in a more nuanced society.

In 2000, a Nigerian man, Rotimi Adebari, arrived in Ireland to seek asylum, having fled Nigeria with his wife and children. A mere seven years later, he made headlines as the first black mayor in the history of the country. Elected as Mayor of Portlaoise, Adebari proclaimed that Ireland was a place of equal opportunity – and isn't that exactly what it should be? Just as Irish immigrants went in droves to the US to make great things happen in 'the land of opportunity', so too does Ireland have the potential to provide prospects for a wealth of individuals looking to make a life in a nation that is democratic, safe and alive with possibility. A nation that doesn't see diversity as a threat to its 'Irishness' but an opportunity to make Ireland a better place to live for a vast array of nationalities.

The critical lack of language skills among Ireland's workforce has emerged in recent years as being indicative of a significant shortfall in Ireland's education system. This skills gap has been a driving force behind the need for Ireland-based companies to look to foreign shores in order to fill their talent pools, particularly in German, Nordic, Dutch and Arabic roles, and the time has come for Ireland to up its game. While I would not for a moment disregard the importance of nourishing Irish language and culture, I do, however, think that the imparting of language skills from a very early academic stage is vital to the prosperity of both Ireland and its indigenous workforce. The ability to speak a language other than your mother tongue not only opens up further growth opportunities and prospects for one's career but also enables one to broaden personal and professional networks.

A society that nurtures the existence of multiple dialects and languages is a society that provides an ideal ecosystem for the growth of multiculturalism. Having bilingualism as the very basic of requirements to achieve a second level qualification would have a horizon-broadening domino effect, not only on the young workforce but the country at large. To supplement school curricula, the establishment and promotion of integration initiatives that spur curiosity and intercultural exchanges would also serve to encourage bi- and multi-lingualism as the future status quo.

The wealth of lessons to be learned from something as simple as exposure to a new culture – an education that was once exclusive to the well-travelled – is invaluable to every nation. With different cultures come different codes of respect and perceptions of appropriate conduct, and the ability to navigate these intercultural webs with ease brings the opportunity to interact and communicate effectively on both a personal and professional level. From an economic perspective, success in the global marketplace requires a truly global outlook. By promoting cultural literacy from the nascent years of one's education, Ireland would foster a cosmopolitan workforce capable of fluidly navigating international trade zones with panache and authority, ultimately extending the already impressive reach and growth potential even further into traditional and new territories. If Ireland were to embrace this, we could very well have the potential to become a case study on the effective management and development of global business relations.

However, an education system intertwined with a unilateral religious focus would hold this process back. Living in South Africa with a rather eclectic group of friends, I was exposed to a plethora of religions on a regular basis, from Judaism and Hinduism through to Islam and Christian Zionism. I often went to synagogue with my best friend Ashleigh, joining her family for Shabat on Friday nights. As a young artist, I attended a Shembe celebration, capturing incredible imagery of the ceremony that ultimately served as the inspiration for my final year artist's portfolio. During Diwali, I'd feast on the barfi in the sweet meat boxes my Hindi friends brought to school and marvel as the fireworks illuminated Durban's skyline. As dusk settled over Durban each day, I'd hear the Sheikhs from the mosque near my school call the Muslim community to prayer. It never occurred to me that this was a unique experience. To me, it was a life that warmly welcomed me to sample religions other than that of my Roman Catholic upbringing.

In my vision for Ireland, students are encouraged to think critically and independently through a blend of theology and philosophy, while simultaneously ensuring that the spiritual background of each student is equally respected. The group of schools under the Educate Together umbrella is already leading the way in this regard by striving to cultivate a critical knowledge, understanding and awareness of the teachings of religious and non-theistic belief systems, unpacking how these systems relate to our shared human experience. Children no longer have to feel isolated for their religious beliefs within a school environment, like the non-Catholic child feeling left out as her class takes their First Holy Communion or the Muslim student who has to request a venue and break from class in order to pray. If schools were originally created to educate and equip individuals to be effective citizens, then promoting cultural and religious equality is a significant move in the right direction.

Celebrating a national holiday that you don't 'own' is, in some ways, like giving a high-five to the nation to which the day of celebration belongs, a nod of approval that says, 'Hey, we like the way you do things.' 'Paddy's Day' is a prime example of how a national holiday can resonate beyond the diaspora to become an international festival of one nation's identity, celebrated by millions of people who haven't so much as set a foot in Ireland.

The adoption of cultural holidays is something South Africans do with gusto. The Holi festival was recently celebrated in South Africa, bringing hundreds of non-Hindu people together to celebrate the change of season in what was a colourful explosion of a street party. Despite the fact that the Holi Festival is celebrated in response to the arrival of Spring, an assortment of South Africans decided there was no problem switching things up to welcome Autumn in the southern hemisphere instead. Some may see this as bastardising a sacred cultural celebration, but I prefer to see it as an opportunity to invite people to experience your culture at its best. Just as the world has opened its arms to a day that celebrates all things Irish, shouldn't Ireland open its arms to the rest of the world in return?

A Festival of World Cultures takes places in Dun Laoghaire each year, attracting over 200,000 people to the small harbour town to revel in a smorgasbord of international song, dance and food; and celebrations such as Nigerian Independence Day, Chinese New Year and Diwali are beginning to take root on Irish soil. These are all a fantastic step in the right direction. I envisage an Ireland where all immigrant communities feel comfortable and welcome enough to share their national celebrations, inviting people to truly revel in their cultural holidays, opening their world up for everyone to have a browse and learn a little more about other nations and their traditions.

And, of course, everyone's favourite part: the food. People often ask me what the national dish of South Africa is and, to be honest, there really isn't one. Instead, we glean culinary inspiration from Portugal, France, Africa, Italy, Greece, Israel and Asia. Sometimes we 'braai'; sometimes we have bacon and cabbage. (Well, in our household, anyway.)

I can vividly remember going to the Durban Indian markets as a child, wrapping myself in a jewel-toned rainbow of fabrics, adorning my tiny arms with bangles that jingled, and munching on spicy potato samosas, not knowing how to abate the tingle the chilli had raged in my taste buds. I also recall paying visits to Giorgos, the Greek man who owned a bakery close to my childhood home, who'd hand me samples of sweet, honeyed baklava or salty spanakopitas as he chatted expressively with my father. An Ireland that dines on the gastronomic delights of countries far and wide is an Ireland that will develop a refined palate primed for culinary discovery.

While we certainly eat out frequently and are somewhat privy to experimenting with flavours from all four corners, I believe we should go beyond the traditional 'dining experience' and plunge into the foodie findings inherent in the authentic street food from nations all over our globe. Imagine if, each year, cities and towns across Ireland welcomed an international street feast where all communities had the opportunity to share their cuisine with the nation. Not only would this allow us to munch our way across continental divides but it would also enable the opportunity to connect with one another through the soul inherent within each nation's edible offerings.

With every good feast comes the need for celebration and revelry. Traditional Irish music is admired the world over and year upon year attracts hundreds of people to our shores.

This rich cultural heritage positions Ireland as a superb stage for traditional musical greats from all over the world. Picture Mali's most popular xalam player performing alongside India's foremost sitar star or Slovenia's leading accordionist in a festival of world music. This could have the potential to market the land of a hundred thousand welcomes as a cultural hub, attracting tourists who may otherwise have never put a trip to Ireland on their wish list.

Could this be the Optimist's Guide to Multiculturalism? Perhaps. But what's wrong with a bit of optimism when it comes to the creation of a vision? When I was younger, my best friend was a girl called Sanushka Moodley. Sanushka was a South African Hindu and she and I became friends in post-apartheid South Africa, a relationship that would not have been possible one decade previously. I vividly recall sitting in the bouncy backseat of her Mum's car, springing up and down on our tiny little bums to Michael Jackson's *Dangerous* album. I smile now, thinking about our innocence as we sang to one another with huge conviction: '*But if you're thinking about my baby/it don't matter if you're black or white*'. The singing was accentuated with the not-so-subtle choreography of me pointing on cue at Sanushka when Michael said 'black' and she pointing at me when he said 'white'. Only a few years previously, *that* would have been an impossible vision.

Initiatives such as One City One People, spearheaded by Dublin City Council, are big steps towards creating an Ireland that respects and embraces diversity, promotes equality and executes a zero-tolerance attitude towards discrimination of any kind. That's the kind of Ireland I want to live in: an Ireland that could teach the rest of the world a thing or two.

The concept of 'Irishness' is going to be different in twenty years – and that's ok. One only needs to look at the strength of the Irish diaspora flung across this planet of ours to know that being Irish has nothing to do with geographical location. Your national identifiers are etched into your very bones. 'Irishness', as it happens, travels rather well. I was always baffled when Irish relatives would enquire when I was 'coming home' despite the fact that I was born and raised in South Africa. After six years here, I finally get it. To them, my mother's departure from Ireland and the subsequent birth of a certain half-Irish, half-French South African didn't really hold all that much weight at all. In their eyes, this homeless mutt never left.

Chapter 14

Roslyn Steer

Music

Roslyn completed a BA in Music and Irish in 2011 and an MA in Music and Cultural History the following year – both at University College Cork and both with first class honours. She was awarded the University College Scholar of Music in 2009 and the Title of College Scholar in 2011. She won the Modern Cultural Studies category of The Undergraduate Awards in 2011 but was also highly commended in the Celtic Studies and Irish category in the same year. She has been a choral scholar at St Fin Barre's Cathedral in Cork since 2009 and she is a composer and performer in a variety of other contexts, mainly as a bassist with the band Saint Yorda.

Music is a nebulous beast, which is what makes it so fun and interesting. Still, despite spending most of my time listening to, creating and writing about music, I don't even know the half of it. Music exists in a baffling variety of contexts in Ireland, each with its own set of needs and desires. A music therapist devising a programme for a nursing home in Galway will probably have a different vision for music in Ireland from a doom metal band working on an album in Limerick. Performers, composers, curators, promoters and audiences all experience music in their own ways and will therefore have different, often conflicting, priorities. The challenge to create a 'vision for Ireland' – a music utopia – is already running into a bit of trouble. It will be a utopia for whom?

A utopia for me, obviously. I've taken this as an opportunity to be completely biased, subjective, demanding and unreasonable – a chance that doesn't come along very often. I am going to be a fairy godmother with the power to create my own personal utopia.

But I am going to need some help with this, so I've done what any self-respecting artist would do and set about trying to steal other people's ideas; asking my peers how they would make Ireland a better place for music. I was met with a deep sense of frustration, especially from those engaged in the music-making side of things on a daily basis. (In a way, this is actually a positive thing: if the next generation of artists is not burning with frustration, there's probably something wrong. We can at least know that the young musicians of Ireland are passionate about what they do.) Luckily for me, my peers also had some helpful notions about how things might be improved.

In my utopia, music would be accessible to everybody, starting from the very young. Music education would begin in primary schools with an emphasis on exploration rather than examination. Primary school children would have the chance to try out music, to explore how things sound and how music-making works. There would be plenty of opportunity to be loud and out of tune. Learning how to read music would become a lived experience – there is no point in knowing the difference between F and F sharp, or between a crotchet and a minim, unless you can hear and *feel* the difference. Reading music would be understood as a tool for creating music, rather than an abstract list of terms, and would be complemented by a range of practical opportunities for music-making. From small classroom ensembles (rock bands, choirs and drum circles) to large orchestras and ensembles formed from multiple schools or community groups, the experience of making music together would be valued as a lasting boon to the development of musical skills and it would also be valued in terms of the importance of listening to others, co-operation, teamwork, hand–eye co-ordination, self-confidence, using your voice and, most important of all, fun.

Learning music *is* fun. So this exciting world of sound would not be closed off to any child just because they can't afford private lessons or they didn't practise their scales for twenty minutes every day this week. Not every child wants to play in an orchestra, but every child who does want to should have the chance to give it a go. Children would pick up instruments early and be allowed the time to develop at their own pace; not an eyebrow would be raised at the fact that they don't, say, have their grade seven by the time they're 10 years old. It wouldn't matter if only a small number of children kept up formal music training: those who stop would do so because they want to do something else, rather than because somebody has told them directly or indirectly that they are 'not good enough'.

The format of secondary school music education programmes would be completely overhauled. The music of dead, white, male composers would not form the bulk of the coursework. The listening components would have a far greater emphasis on practical listening skills instead of learning off lists of facts. Composition would be taught as a creative process, inviting young people to create their own works of art, instead of learning off formulae that 'work'. There would be far more extra-curricular musical activities and resources available to young people and more contexts in which they could build up their confidence at performing.

In the course of my research for this essay, I spoke with peers about the current state of music in Ireland and I discovered that many of us feel the same dissatisfaction towards certain prominent members of the Irish music industry. This is not to say that these people as individuals are causing consternation, it's more that they represent something wrong in the industry: the sordid, veneer-based entertainment that is being flogged as Irish popular music. The style of music these people trade in is not a problem (there are people out there who enjoy it) but it appears to be quite a toxic irritant to a lot of musicians in Ireland. So while the *individuals* will not 'go to Room 101' along with the Junior and Leaving Certificate courses, they might be taken off the air for a while so that we can get a bit of fresh music in. Of course we'll put them back on if they promise to be good.

In my utopia, there would be ample support for all kinds of musicians in Ireland. We would enthusiastically promote and celebrate the wealth and diversity of our musical talent, and not just the chosen few who tick the right boxes as being sufficiently marketable. Music that is made to sell would not necessarily be left out; it's more that everything else would be welcomed in as well. In my utopia, it would be easy to find out about *all* of the wonderful musical activities that go on across the country, not just the same venues hosting the same acts over and over again. We do get glimpses of them now and then, but many of our most talented musicians are creatures rarely seen or heard above ground. It is time to look beyond the pale. In my utopia, there would be funding for new works in every genre, all over the country, particularly for new and emerging projects. And the people involved in these projects wouldn't be expected to have a pre-existing portfolio of wild successes in order to have any chance of getting off the ground.

When you turn on the radio in this utopia (people would still be listening to the radio), you will hear Irish artists being given the air time and attention they truly deserve. Of course, the audience of my utopian nation would be actively seeking new Irish music, but they wouldn't have to wait for a dedicated 'Irish hour' to hear it. This music would be available as part and parcel of the mainstream, prime-time output of Irish radio stations and not just the token few golden oldies and the 'flavour of the month' band (who sound suspiciously like last month's flavour of the month band). Similarly, the Irish press would be out there scouring the cities and the countryside for new music. We don't have a very big country, but we do have a lot of music and an audience somewhat dependent on the media to give them an indication of what is out there.

There would still be physical releases (as in CDs, tapes, vinyl records) because they are lovely things to have and to hold, and independent record stores deserve the business. The proprietors of these places often single-handedly promote and preserve the output of local or lesser-known musicians.

(Try going for a browse some time. It's like going to a fancy wine shop, only the flavours are for your ears.)

Musicians and venues would do their part by really encouraging people to come in and listen, so there would be more gigs and early shows suitable for all ages. Music that might at first seem inaccessible to new listeners by virtue of style or 'scene' would be more inviting to curious ears. It wouldn't cost over €70 to get into any single music event, but every musician would be paid something that recognised the investment of time and money that goes into bringing music to the public. There would be a balance and a sense of mutual appreciation for the time and hard labour of everybody involved in bringing musical projects, large or small, to a state of fruition.

Online access to music and musical resources is and would continue to be an important part of music consumption in Ireland. And the internet would be a vital tool not only in the promotion of music events and new releases, but also for creating more open systems of communication between music-makers in Ireland. Musicians would know each other and help each other; young composers would talk to one another and find musicians to experiment with their ideas. Young bands would listen to one another, be willing to help out one another or give advice. We are in a risky business but an attitude of friendliness and consideration would make sure that music stayed a fun and positive experience for everyone involved.

In my utopian reverie, I am somewhat ignoring such practicalities as economics, politics and the harsh realities of getting your music heard. But, you see, I really don't think the first thing an aspiring young musician or composer should hear is how awful their chosen profession is; that people are mean and nobody will play your music or let you use their venue unless you're going to sell a hundred tickets and two hundred pints. Being constantly on your guard makes for tetchy working relationships, which is a scourge on creative collaborations that might otherwise flourish into something really beautiful. There is a difference between having your head screwed on and being needlessly defeatist or aggressive.

In my utopia, it would never be forgotten that music is fun. It is hard work, certainly – but heartache always comes with something that you care about. The 'job' part of being a musician would not be so complicated in this utopia of mine, since all musicians would be paid fairly for the hard work they put in and the expenses they incur. Musicians would not all be millionaires but they would be able to pay for their equipment, their rehearsal time, their travel costs and maybe have a few bob to invest in the next gig or the next stage of their development.

There would be enough resources and opportunities for the young people of Ireland to make the music they want to make, whether that means being welcomed into the local trad group with no more than a bodhrán and a dream; seeing your composition for twelve cellos and a percussion ensemble being performed; getting your band from A to B; having a classic hits night in the local nursing home; or holding a hip hop night in the local primary school. It wouldn't really matter what the project was: anybody who wanted to create some music would be afforded the opportunity to do so. There would be funding for large-scale, ambitious projects and local resources available to anybody to use. There would be a greater number of music venues that were designed and managed specifically for musical performance. These are not necessarily safe decisions, but playing it safe is an awfully stifling burden to bear. Not everything would be outrageous, drastic and experimental (although, there'd be a bit of that) but the playing-it-safe option wouldn't be the default. Music is an ideal medium in which you can *not play it safe*: you can try out an idea, create a world, be a different person and see what kind of sound your idea makes.

Maybe – and here comes the really subjective part – a site-specific staging of Berg's *Wozzeck* would be a real option for an opera theatre company. And maybe there deserves to be a fairer proportion of acts from outside of Dublin in the Hard Working Class Heroes festival. And maybe Arts Council funding would be extensively advertised and evenly distributed across various musical genres. But hey, I'm a fairy in charge of a utopia, so a flash of the wand and we can have both the safe *and* the unsafe productions: funding for everybody and all the music of the country on a level playing field. Hooray!

Fortunately – now here's the serious part – many aspects of my idealistic vision *already* exist in Ireland. We have an astounding wealth of diverse musical talent that doesn't need a fairy godmother with a magic wand – just a little more energy and exposure. There are some radio programmes that do an excellent job at promoting a variety of Irish music. The Arts Council does fund some very worthwhile projects. And there are schools and youth projects that give great encouragement to musical education. The bones of things, I think, are here already.

The most burning and urgent desire that I have for music in this country is that everybody should have the chance to enjoy it as much as I do. The fact that I can share and experience so much great music every week is a truly fantastic thing and, as I mentioned before, I don't even know the half of what goes on. I would urge and advise that you get out and explore your local record store, stop and look at the posters that advertise music events, check the listings, search the blogs, find something you might enjoy – and check it out. If you like what you hear, consider supporting the people that brought you the music and remember to share it with others. Music is good for you – and Ireland has a lot to offer.

Chapter 15

Eimhin Walsh

Religion

Eimhin was a foundation scholar at Trinity College Dublin and graduated with a gold medal before completing his postgraduate studies in Theology at the University of Oxford. He was the inaugural winner of the Religions and Theology category of The Undergraduate Awards in 2009. He is currently back at Trinity, working towards a doctorate in Medieval Ecclesiastical History. He has also worked extensively within the NGO sector and has served on the boards of various charities, including Amnesty International in Ireland.

The history of Ireland cannot be understood apart from the history of religion in these islands. Druids, hermits and priests have left an indelible mark on Irish society through the ages. Even the founding fathers and mothers of modern Ireland acknowledged a political role for the divine 'from whom is all authority' in reflection of the esteem in which religion was held by society. As the embryonic state found its feet, the Roman Catholic Church emerged as the spiritual mouthpiece of the Irish soul. So robust was this force that 'Catholic Ireland' remained immune to the effects of secularism until relatively recently.

Now the religious hegemony has collapsed. Now the state wrestles with secularism. Now religion finds itself relegated to the private sphere. One could easily lament that the history of religion in twenty-first-century Ireland is a sad story of decline. Is there a future for religion in Ireland? What will it look like?

The Catholic Church in Ireland has been beset by scandal after scandal. It is often charged with resistance to change, which has resulted in a shrinking church attendance from 85 per cent in the 1970s to 45 per cent in the 1990s. The numerical decline in religious practice is frequently interpreted as the vindication of Marx's presumption that as society advances religion will deteriorate.

These Irish trends, however, should be set in global context. While European religiosity ostensibly seems to be in decline, across the world religion is growing. Eric Kaufmann's demographic analysis of religious populations has demonstrated that the global increase in religiosity cannot but impinge on the landscape of European piety. While conversion to religion remains a small factor in the growth of religions, by and large most people will inherit their religion from their families. It follows then, that population growth will naturally impact the texture of religion. For example, 97 per cent of population growth occurs in the developing world, where religiosity is stronger than in the west. The UN estimates that by 2050 there will be four African people for every European. Since the most devout part of the world is growing substantially faster than the less devout, the statistics indicate that between 1970 and 2000 the proportion of the global population that professes a religion increased from 81 per cent to 85 per cent. Although growth rates are not as high now, it is still anticipated that by 2025 87 per cent of the world will be religious.

The logical outcome of these global trends will be monumental cultural changes across Europe. Kaufmann believes that the declining birth rate in Europe, coupled with the ageing European population and the exponential growth of the global south will lead the north–south population disparity to peak around 2050. This will drive migration of expanding religious populations into the declining secular west. These trends are already being felt in Ireland. The last census recorded a 73 per cent increase in Pentecostal and Apostolic church affiliation and a 51 per cent increase in Islam compared with a 5 per cent growth in Roman Catholicism and 6 per cent in Anglicanism.

Some will read these trends triumphantly, arguing that the numerical growth signals the recovery of religion. A nuanced reading of the last census cautions against such triumphalism, since the largest growth was the 320 per cent increase in atheism. Debating numerical growth or decline is not a sufficient indicator of the healthiness of religion; rather, one must look deeper into what these statistics mean for the fabric of religion in Ireland. While secularism is undoubtedly increasing at the moment, the major growth within religious communities is amongst the more conservative and fundamentalist expressions of faith. When read alongside the declining population, it would appear that the future of religion in Ireland will be increasingly fundamentalist.

Fertility rates are disproportionately higher amongst fundamental expressions of faith. In the US, fundamentalist Christians tend to have one child more than the national average. Orthodox Jews will typically have 7.5 children while mainstream Jews only have two. Those Muslims that favour Sharia law have twice the number of children than those opposed to it. Despite secularisation, the Europe of the future will be more religious than today and because of the demographic benefit afforded to fundamentalists, religion will be more extreme than today.

How should Ireland respond? One possibility would be a clarion call to all mainstream religious to go forth and multiply. A more sensible approach is to pause and consider the root causes of religious fundamentalism.

The term 'fundamentalist' emerged in the 1920s in response to the challenges that modernity posed to religion. The American Baptist preacher Curtis Lee Laws charged Christians to 'do battle royal for the fundamentals' of Christian belief.[1] It is a reactionary defence tactic from a community that understands itself as besieged. The programmatic rationalism that has undergirded intellectual progress since the Enlightenment is perceived as attacking the fundamental tenets of religious belief. Tied to rationalism was the advancing secularism in western society. One only needs to consult the writings of aggressive scientism in order to find populist rhetoric denouncing faith as futile and pronouncing religion a stumbling block to progress. Secularism is often grounded on the belief that since religion has a smaller role for a smaller number of people, it should be privatised and removed from the public square.

Once-powerful institutions have lost the voices they once had. From the modernist perspective, fundamentalists are reactionary radicals seeking to return to a bygone age. From the religious perspective, modernism is catastrophic and amoral change that tears apart community, social bonds and collective meaning. Sociologist R. Stephen Warner notes that this polarisation of perspectives causes a resurgence of anti-rational religion.

Yet neither the secularist nor fundamentalist perspective holds a monopoly on truth. Maureen Junker-Kenny has written that 'the requirements of reason protect religion (as well as science) from descending into fundamentalism.'[2] It would appear that both modernism and religion have strayed towards a form of universalism that is potentially destructive and leads to expressions of fundamentalism on both sides. The suggestion that in modernity religion becomes a set of individually held beliefs without meaning for the broader society is an assertion that limits the emancipatory potential of cultural pluralism. If society is composed of people who are valued for what they are, and if what they believe contributes to what they are, then these beliefs have got to be considered in the creation of public policy.

The exclusion of religious voices from public reason presumes that religious perspectives can only disturb the peaceful and progressive inclinations of society. Religious beliefs can be enriching resources for society, and can often offer critiques that challenge societies to think deeply before acting. But at the same time, an increasingly fundamentalist religion is more likely to offer a vision of society that is totalising and dogmatic. The solution seems to be predicated on finding a *via media*, and it seems to me that the recovery of reason can be that voice of balance.

But why is balance needed? The modernist vision that birthed secularism is obsolete. The universal rationalism that scientism, secularism and modernism articulate simply does not exist. The positivist worldview that for a thing to be rationally acceptable it must be scientifically verifiable is out of sync with the experience of what is now regarded as postmodernity. The ongoing journey of scientific discovery is revealing the complexity of our understanding of the world. Genetics now converses with epigenetics, and physicists that heretofore examined atoms now concentrate on the sub-atomic. In a postmodern world the mind cannot be exclusively comprehended through neurobiology – the value of human experience must be taken seriously. The world that was demystified by modernism is being re-mystified.

One can see the popularity of the quest for meaning by looking at the ever-expanding mind/body/spirit sections of bookshops. The collective obsession with economics as a predictive (almost omnipotent) force could be read through a quasi-religious lens. In a postmodern world, symbolism, ritual, belief and meaning are regarded as possessing intrinsic value, while paradox is not necessarily to be rejected but can be accepted as part of a worldview that is sceptical of all meta-narratives. Within postmodernity the spiritual once more becomes an authentic conversation partner. If modernist secularism is obsolete, it follows that its antithesis – religious fundamentalism – is likewise obsolete. A middle ground is needed because the universalising claims of secularism and fundamentalism do not speak to the conditions of postmodernity.

Reason can be a mediator in the clash of fundamentalisms when it is used as the yardstick to measure competing claims. For public discourse, the recovery of reason in part means respecting the insight that theological reflection can bring; that religion is not *a priori* irrational. For religious discourse, this means distancing itself from fundamentalism. I cannot speak substantively to the former claim, but it is to the latter that I will direct the remainder of this essay.

Anticipated demographic changes suggest that religion will become more fundamentalist. But traces of fundamentalism can be found even within mainstream religious traditions. The Roman Catholic Church has launched the 'New Evangelisation' with the aim of winning back those traditionally religious countries, like Ireland, that have capitulated to secularisation. Its rhetoric implies a battle for religion against society. Another example is the pressure from conservative lobby groups seeking to explicitly exclude homosexuals from leadership positions within the Church of Ireland.

While it is regrettable to see these arguments in a tradition that was founded on the premise that opposing theological perspectives should be held in creative tension, it is even more concerning to consider the reasons given to substantiate the conservative position. The three that top the list are: that the bible condemns homosexual practice; the tradition of the church has always been against it; and that society is forcing the church to change. In sum, the reasons highlight a reluctance to embrace change and an antagonism to the changing cultural values of society. This reveals a monumental ignorance of the history of religion.

The challenge of recovering reason will mean for conservative Christians that they must critically re-examine and re-appropriate the fundamentals of their faith. I believe that this process must begin with a reconsideration of the history of religion. The church historian Henry Chadwick once said that 'nothing is sadder than someone who has lost his memory, and the church which has lost its memory is in the same state of senility.'[3] Religious fundamentalists seem to have acquired a cultural amnesia because the history of Christianity (and many other religions) is a history of *change*.

While it is often alleged that religion is fixated on the past, it is more accurate to say that religion is fixated on a particular view of the past that paints history as monochrome. It is the duty of church historians to challenge religion to keep its memory alive and interested in the complexity of its own story. As Milan Hubl remarked, 'The first step in liquidating a people is to erase its memory.'[4] Our culture is such that we have privileged the present at the expense of the past, but it is only after mature reflection on events past that we can really comprehend how we ought to act in times present.

The critical examination of history lifts religious fundamentalists out of the heat of the battle and exposes them to the development of their traditions. It helps to frame perspectives and to contextualise doctrine; it is thus an enriching and enlightening pursuit. Yet there remains a sense of fear that critical awareness may cause a house of cards to tumble. That is the challenge that fundamentalists must be encouraged to overcome if they are to rescue religion from irrationality.

John Henry Newman, founder of University College Dublin and a scholar of the early church, observed that to live is to change and to be perfect is to have changed often.

He was writing about the development of Christian doctrine and he understood that a critical reading of Christian history revealed the gradual formation of theology. Doctrines, theological opinions and ethical perspectives, while perhaps deriving from divine revelation, are always mediated to people through culture. Consequently, all exponents of any theological or ethical premise should attempt to understand the cultures that have preserved them.

Many of the current theological battlegrounds centre on historical developments: the prohibition on women's ministry, married clergy and condemnations of homosexuality are all relatively late developments in theological history. And yet religion consistently denies this. Newman's aphorism highlights the necessity of critical evaluation of the past in order to comprehend what is actually fundamental and what is incidental. Newman's example also shows that such critical engagement does not have to weaken personal faith.

As we look over the history of religion in Ireland we can see the richness of diversity. This is the island that produced Christian philosophers John Scotus Eriugena and Bishop George Berkeley, the explorer Saint Brendan, the evangelist and founder of the Plymouth Brethren J.N. Darby, the literary genius Dean Jonathan Swift, the ascetic Saint Kevin, the biblical scholar Archbishop James Ussher, the allegorist C.S. Lewis, and the visionary nun Mother Catherine McAuley. Each was brilliant in their own way and each created their own unique interpretation of faith.

Religion is a patchwork quilt made up of various perspectives. Sadly, religions have become so embattled in fighting secularism that they have become institutionalised and closed themselves off to the richness of this diverse tradition. An openness to change creates in religion an exciting impetus that gives a past and future dimension in addition to a present focus. It removes religion from the rut of present battles and enables it to focus on what it can become. Empowered by the knowledge that change is not a break with tradition but part of it, religions can discard the flat one-dimensional versions they currently espouse and re-engage the creative imagination. Confident in the integrity of their values, they can offer a radical vision of society, equipping its members in their individual pursuit of meaning.

Ireland has an opportunity to highlight to the rest of the world how to deal with these changes facing religion. A mature solution rests on renouncing two fundamentalisms. Aggressive secularist policies should not be pursued by the state, which should instead welcome the voice of religion in the conversations that build the future. The bigger questions, however, are for religions themselves. In order for religion to be regarded as a legitimate partner, it must distance itself from fundamentalism. The rapid expansion of fundamentalism, coupled with anticipated demographic changes and current institutionalisation, means moderate religion is haemorrhaging. By 2050, anti-rationalist fundamentalist mentalities will be so deeply ingrained that they will be difficult to shift and religion will lose the voice it deserves.

If it is to guarantee its future, religion must recover reason by embracing a theology that is open to social change. As they look (sometimes with trepidation) for guidance in responding to change, I suggest that religions should look to their past.

Notes

1 Packer, J.I. (1958), *Fundamentalism and the Word of God.* Michigan: Inter-varsity Press, p.29.

2 Junker-Kenny, M. (2011), 'Witnessing or Mutual Translation? Religion and the Requirements of Reason', in L. Hogan, S. Lefebvre, N. Hintersteiner and F. Wilfred (eds.), *Concilium 2011/1: From World Mission to Inter-Religious Witness.* London: SCM Press, p.112.

3 'The Very Reverend Professor Henry Chadwick: Obituary', *The Times*, 19 June 2008, p.64.

4 Milan Hubl quoted in Kundera, M. (1983), *The Book of Laughter and Forgetting.* Harmondsworth: Penguin, p.159.

Further Reading

Coghlan, N. (2010), 'So Does Religion Have a Future?', *Studies*, 99.

Kaufmann, E. (2010), 'Shall the Religious Inherit the Earth?', *Studies*, 99.

Warner, R.S. (1993), 'Work in Progress Toward a New Paradigm for the Sociological Study of Religion', *American Journal of Sociology*, 98.

Chapter 16

Dr Fred A. English

Research

Fred is a physician at Dublin's university hospitals and a maternal health researcher at the Irish Centre for Fetal and Neonatal Translational Research. A former fire-fighter, Fred was educated at Institute of Technology Sligo and University College Cork, and won the inaugural Medical Sciences category of The Undergraduate Awards in 2009. Today, his research focuses on pregnancy-specific illnesses that affect expectant mothers and their babies. His work is the product of worldwide collaboration and has featured in leading medical journals.

Impossible Is Nothing. Just Do It. Is féidir linn. Such slogans have become the mantra of young people today. For all the obvious damage it did, I believe the Celtic Tiger period and its aftermath have instilled courage in young Irish people; a courage to take on the world and to see all as equal.

The last ten years have left a memory that forces us to base confidence on an accurate appraisal of our skills and weaknesses on a global level. No longer are we evaluated by the amount of cash in the country. There now exists a resolve never before experienced in a young Irish population. This is a competitive generation for whom nothing is impossible. A generation with little interest in a job for life, but instead, a burning desire to leave a mark on the world. Today's twenty-somethings are fortunate to live in a time in which the pursuit of knowledge and innovation are fashionable pastimes. Our small country has never been so well placed to enable its young people to leave the world better than they found it, particularly through involvement in quality research.

Ireland is a nation playing to its strengths in a concerted effort to become a world leading force in the quest for understanding and betterment. In the 1980s, Ireland's research impact was equivalent to that of Bangladesh. By 2000 we had progressed, and today our output places us among the top twenty countries in the world in terms of research. So the challenge in the future will not be catching up but *keeping up*. This essay is an insight into some of the ways in which Ireland is driving competitiveness and growth in research and innovation, offering a snapshot of now and a hopeful glimpse into the not-too-distant future.

PLAYING TO OUR STRENGTHS

Ireland has a proud history in curiosity. The lives of great names like Yeats, Shaw, Joyce, Walton, Beckett and Boole are chronicled on plaques that dot the streets of many towns. We have an international reputation as a considered and thinking people who solve problems better than most.

We have a unique reverence for knowledge and curiosity and it is ingrained in our identity. This is reflected in our contribution to the humanities. The use of allegory and the promotion of kinship have long been appreciated by Irish people. Saint Patrick had us hooked at the mention of a shamrock. Whether through the intricacy of the Book of Kells or Seamus Heaney's considered poetic analysis of the Troubles, for generations Irish people have been connected with the study of self.

Young Irish people today understand that all knowledge is valuable and that cultural achievement is equally as valuable as commercial advancements. The humanities offer society an opportunity to understand and engage with the forces that have moulded it in the past and continue to shape it. Our young researchers are encouraged to examine our unique cultural context with the belief that quality of life can improve simply through better understanding of our place in the world.

In the future we will continue to look inward, honestly examining our own culture and striving to create a more inclusive and transparent society. Our potential to export the lessons we have learned from the past will also be realised. Having found peace on an island harbouring many creeds and cultures, Ireland is in a unique position. There is no more noble export than peace and diplomacy. Peace, as an industry, is an area in which Ireland has the capacity to change the world. Leaders in this industry, including Mary Robinson, John Hume and David Trimble, have already exported Irish lessons to foreign shores. The future will see Ireland continuing to face hostility and disagreement in an effort to bring peace and reconciliation to our time.

If one could identify a single national strength that unites all others, an asset that bridges our past and present, it would be that which stems from who we are and how we do things. The Irish psyche is resilient, flexible and willing. Irish people are known for our communication skills and leadership qualities. Our strength in this regard has placed Irish researchers in the top ranks of many of the most prestigious research consortia. These positions enable the representation of Irish interests worldwide. Our wide diaspora is a unique asset and has raised awareness of the Irish work ethic and its positive energy. It is widely accepted that our approach to problem-solving is different from that of other nationalities. We are a welcoming and accepting people. As the world continues to become a smaller place, large, innovative, companies and the brightest minds will look to invest in a place where doing what they love is easy, where people will listen to what they have to say and where life is enjoyable for their families. That place is Ireland. *Cead míle fáilte* is more than just a postcard tagline.

Often the relationship between science and the humanities has been portrayed as akin to cousins who limit amicable chat to weddings and other such infrequent family gatherings. The humanities have been labelled as retrospective and science as experimental. However, the rapidly changing pace of advancement is making both increasingly practical.

For the first time in human history discovery is out-pacing application. Today, research progress is being limited by the lack of uses for our discoveries. This quest for usefulness is pulling the arts and the sciences together, with world-leading results right here on our doorstep.

A prime example of the aforementioned collaboration rests in our pockets. Ireland is quickly becoming the development capital for smartphone applications. Whether it's the brainchild of a fledgling company run from a bedroom or that of a multinational in Silicon Valley, Ireland is the place to get your application into pockets around the world. Advances in psychology, behavioural science, engineering and mathematics mean that, not only do we know more about what people want, we're also better at delivering it. Our booming creative media industry is the product of skilled programmers, imaginative social scientists and astute business brains. Creative media is constantly applying knowledge to everyday situations. Ireland has proven to be the ideal site for elegant design, evaluation, modification and ultimate marketing of an application to the world. Our unique size and wide knowledge base makes the production process efficient and effective. It is not surprising that web giants choose Ireland when they want to run their company on this side of the world. In the future, Ireland can build on this early success and become a greater force in the creative media industry, with more home-grown companies delivering top-rated apps. Don't be surprised when the next life-easing app bears the tag 'Made In Ireland'.

Ninety per cent of the data in the world has been created in the last two years. Most of it sits unstructured and awaiting interpretation. This mass of potential knowledge has created the field of so-called 'big data' and provides an opportunity for Ireland to excel. We are becoming increasingly efficient at recording the world around us. Be it the growth velocity of grassland, the impact of global warming, consumer preferences or the effect of a particular disease on a population, every element can now be recorded, resulting in these large and complex databases.

Given the rapid pace of data generation, there exists a shortfall of information scientists, i.e. the people who might sift through this mountain of numbers. Moves are afoot to increase the number of Irish graduates in this area, not specifically to fill positions but to develop our understanding of such data and find a use for it. Ultimately, Ireland hopes to become the world leader in creating tools for analysing big data platforms and in creating models around identifying patterns. Ireland is also vying for supremacy in the ancillary areas such as data protection and information security. The recently announced multi-site INSIGHT research centre is the ideal foundation for Ireland to begin its journey to the top of the data field. From farming to pharmaceuticals, taming big data is the next bastion in technological advancement and, thankfully, Ireland is right where it needs to be.

Research in maternal and child health is an area undervalued in many cultures, but Ireland has an excellent track record in this field. Irish people appreciate and understand the importance of a child's healthy growth and development; over the last hundred years, Irish researchers have made significant contributions to the field. The Dublin Method, a method developed in 1915 and still used in labour today, has made childbirth safer worldwide, protecting the lives of countless mothers and their babies. Irish child health researchers have also achieved great impact through the development of many methods to monitor infant and child nutrition in the developing world, helping millions of children enduring famine and deprivation.

These achievements have provided a foundation for the future. Ireland is set to become a world leader in the early detection and treatment of some of the most serious diseases that occur around pregnancy. This year's establishment of the Irish Centre for Fetal and Neonatal Translational Research (INFANT) will see our top perinatal researchers come together under one roof to further our understanding of the conditions that take the lives of women and their babies. This institute will work with industry to develop tools for diagnosis and treatment of such conditions. Irish efforts to make pregnancy and early life safer are intensifying and are certain to yield results that will span the world.

A deep connection with the land has long been part of the Irish identity. Whether through one's own experience, shared stories with relatives from rural backgrounds or the exploits of John B. Keane or Miley Byrne, every Irish person understands that working the land is part of who we are. Over the last twenty years, Ireland has become a world-leading force in agriculture and our methodology is quickly becoming best practice around the globe. The ways in which we think about agriculture are changing and the tentacles of technology are beginning to extend to every aspect of the farm. Food science is a rapidly expanding area. Progress in key developments and ever-increasing collaboration will strengthen our position and assure our superiority as experts in the process that takes food from farm to fork.

Many of our forty shades of green are comprised of rolling fields of glistening grass. To most, grass is a simple and ubiquitous plant – but to the researchers at Teagasc, our Agriculture and Food Development Authority, grass is a green blanket of potential. Grass could be seen as the nexus of all we're good at. This humble plant is the staple diet for farm animals who, like humans, are what they eat. Simply put, Ireland produces the best grass in the world. Our ability to produce world-class grass is an unlikely result of our capacity in plant genetics, farming methods and the aforementioned big data. Our proficiencies in these areas enable the creation of high-quality raw materials and the close monitoring of growth and development. This success is owed to a simple cyclical process, involving farmers, scientists and industry, which is in fact a model to the world. Ireland is already top of the pile in this area, yet the future is full of opportunity. As the population of the world expands, so too will the demand for food and its ingredients. Our researchers will move from experts to gurus, guiding emerging economies to produce better, healthier foods.

Precision agriculture is the concept of managing an entire farm as one entity with the goal of optimising returns, preserving resources and delivering the finest-quality end product into the food chain. Increasingly, data is guiding the day-to-day decisions Irish farmers make. As one of the first countries in the world to breed cattle on the basis of their genetics, Ireland has a unique advantage in this area. Genetics is being used to predict traits in animals and to guide breeding practices. This precision extends into crop production, where genetics and growth data can be used to produce prime foods with utmost efficiency and effectiveness. Our long history on the land and our expertise in precision farming has led to both the development of indigenous industry and the attraction of overseas interests. Researchers at University College Cork's Alimentary Pharmabiotic Centre (APC) and Teagasc are at the interface between agriculture, food and medicine. APC scientists link the finest aspects of food science with the strength of the food industry to create unique health-maintaining foods. The APC and other Irish collaborative projects such as Food for Health Ireland are striving to create foods based on knowledge. Such customised nutrition will result in better food for all and bespoke diets for patients with illness and people at the extremes of life, such as infants and older adults.

The potential for the Irish agri-food industry is boundless. Indigenous industry continues to grow in order to maximise quality raw materials. Our unique agri-science capacity is creating sustainable employment and worthwhile products. With this growth continuing into the future, Irish labels will quickly become a more regular feature on the shelves of the world.

Irish people are more likely to discuss the weather than new discoveries in genetics or how an app made in Cork is making waves around the world. However, a change may be afoot. The powers that be are hopeful that science can become more accessible and familiar to the public. While the public purse funds much of our output, communication skills remain a recognised deficit in the scientific community. However, the culture is shifting to one where researchers want to engage with the public. It is hoped that social media can work with this change in attitude and create a discourse; people will begin *liking* and *commenting* on research that might be *trending*! Social media, coupled with plain language explanation, has the power to bring complex topics out of the laboratory and into to local pub. In the future, the Irish public will not only have a greater understanding of research but they will be empowered to appraise work and question its relevance. Therefore, they will feed into the system themselves. Who knows, we may even see a new breed of science celebrities or a research-orientated reality TV show.

Be it through peace, creative media, maternal health, farming or food, Ireland's creative potential is yet to be fully realised. What we already know is forcing us to think about how to use this knowledge. Ireland is a country that is closing the gap between discovery and application. Research and innovation offer short-, medium- and long-term solutions to the problems Ireland faces at present. Our strong focus on knowledge-creating research in all disciplines and our increasing emphasis on impact and application will yield benefits that will last for lifetimes to come.

Our assets range from our unique geography to our charismatic people. We are a small and nimble country for whom nothing is impossible. What tomorrow holds is never certain but one thing is definite: the future is bright in Ireland.

Acknowledgments

Fred would like to extend thanks to Professor Mark Ferguson, CSA/SFI; Dr Eucharia Meehan at the HEA and IRC; Dr Frank O'Mara at Teagasc; and Professor Louise Kenny at INFANT.

Chapter 17

Rachel Carey

Road Safety

Rachel is a PhD *student at* NUI *Galway, where her research examines the impact of road safety advertising campaigns on driving behaviour. She recently received the inaugural Rothengatter Award for the best student presentation at the fifth International Conference on Traffic and Transport Psychology. She is an* ad hoc *reviewer for the* Journal of the Australasian College of Road Safety *and she lectures in Social Psychology at* NUI *Galway. Rachel won the Life Sciences Category of The Undergraduate Awards in 2010 for her essay on the study of time in co-ordinated movement.*

Road safety is an issue that affects every corner of Irish society, provoking an emotional response, a heated debate and a myriad of empirical research responses. Researchers and practitioners of road safety aim to reduce the number of preventable deaths on Irish roads by identifying the reasons behind their occurrence. We are not infallible, and some road traffic collisions cannot be avoided. However, human factors account for a majority of collisions, and small changes in driving habits can help to prevent injury and to save lives. As the number of national and international bodies dedicated to road safety grows, and as increasing amounts of resources are invested in this area, it seems timely to examine what the future holds for road users in Ireland and elsewhere.

After completing my undergraduate degree, I decided to undertake a PhD (funded by the Irish Road Safety Authority) in social psychology, with a specific focus on driver risk-taking. Since then, I have spent three years conducting research in this area, looking at ways to improve road safety and examining the effectiveness of road safety advertising campaigns on speeding and other types of risky driving.

Like many a PhD student, for the last three years I have become almost entirely absorbed by, and dedicated to, the topic I am studying. This type of research is not something that is easily left behind at the end of the working day. More often than not, I spend 'days off' tearing out newspaper articles, bookmarking websites and recording television programmes that relate to road safety. Countless days, after travelling home from the office I have ended up sitting in the car outside my house, glued to a radio programme that is sometimes only vaguely relevant to my research. Countless nights, I have woken up with a thought about something I've read, or with an idea for another experiment (all of which are invariably forgotten by morning).

I describe these experiences to give some context to the investment I have made in this topic. As I come towards the end of this particular programme of research, I find myself increasingly conscious of, and curious about, the future of research in this area. I wonder about the bigger picture. What are the next ten, twenty, fifty years likely to hold for road safety research, policy and practice? And what is it that I would like to see happen?

These are not straightforward questions. While there is little debate surrounding the changes people would *like* to see occurring in road safety in the coming years, there is no consensus about how this is likely to be achieved. Specifically, road safety practitioners aim to make the roads safer for all road users, working towards a future where no lives are taken as a result of preventable accidents on our roads. It is in choosing the strategy that will accomplish this aim that people differ, and this is where road safety becomes a contentious issue.

Discussions about road safety in general, and driving in particular, tend to draw on a number of approaches, including psychology, engineering, sociology and education. Since my background is in social psychology, I usually adopt a psychological perspective. Before attempting to modify behaviour, psychologists attempt to understand why it is occurring in the first place, and whether we can expect it to happen again. Some psychologists would suggest that the best predictor of future behaviour is past behaviour. With that in mind, a good starting point might be to examine recent changes in Irish and international road safety trends.

In Ireland, as in most other developed countries, the statistics are increasingly reassuring. The number of deaths from road traffic collisions in Ireland for 2012 fell to 161, the lowest level on record. If current trends continue, we are likely to see a further decrease in fatalities resulting from road traffic collisions in high-income countries over the next few years.

So what is causing the reduction in road traffic fatalities? There are a number of factors at play, and these will form an important part of future policy and practice. Proposed factors include more evidence-based road safety campaigns, increased education in schools, improved road infrastructure, severe punishments for risky driving behaviours and greater policing. The issues correspond to the so-called 'pillars' of road safety strategies, the *three E's* of education, engineering and enforcement. Each of these has a distinct role to play in the coming years, though the nature of the interplay between them remains a source of debate among researchers.

EDUCATION

Educating people about important road safety messages forms a core part of short-term and long-term road safety strategies. Education approaches involve school-based campaigns, campaigns at a societal level, as well as driver education and training. National road safety authorities are increasingly providing teaching resources to schools and colleges, and promoting national road safety weeks to raise awareness about and encourage local efforts to improve safety and to reduce road traffic fatalities. Education-based approaches make intuitive sense. They are popular and, at least anecdotally, have a positive impact. My five-year-old nephew, for example, has been known to reprimand his parents for not looking both ways before they cross the road.

In recent years, the mode of presentation of these types of approaches has changed. In driver training, technological advances are leading to the inclusion of in-vehicle data recorders that monitor driving behaviour. A number of these types of devices are already in use, some of which are connected with fleet drivers and insurance companies. The idea is a simple one: driving (e.g. your speed, braking distance and lateral position on the road) is continuously monitored by this device and a driving summary is provided on a daily or weekly basis. Managers of fleet drivers examine this information in an attempt to promote safe driving among employees, and insurance companies use the data to determine insurance premiums.

Previously restricted to fairly bulky, expensive data-recording devices, the future for this area is bright and may involve the use of phone apps. A number of similar apps that record and reward safe driving behaviour are already in development and may lead to reduced insurance rates. In an economic climate where more and more people are struggling to find money to keep their cars on the road, and little savings mean a lot, this approach may prove popular and rewarding.

Technological advances have also led to new approaches in school-based education campaigns, including the use of simulated driving experiences and other forms of experience-based learning. In Italian school-based campaigns, for example, the police are beginning to include live crash videos, usually depicting a road traffic collision recorded from roadside cameras. Given the increasing popularity of interactive media and children's mounting enthusiasm for, and ability to use, technology, this may be an important aspect to focus on in the future.

Despite their popularity and pervasiveness, road safety education programmes delivered in schools have, to date, largely not been based on empirical evidence. Dr Frank McKenna, keynote speaker at the 2012 International Conference of Traffic and Transport Psychology, pointed out that while education strategies in road safety are appealing, their effectiveness has not always been supported by empirical research.

The key point here is: in order for education campaigns to be effective, they need to be designed in accordance with theory and evidence. Looking forward, the future for road safety education programmes is in identifying and implementing education strategies that are evidence-based, and evaluating them on a regular basis.

Most researchers would suggest that education, by itself, cannot create the large-scale changes needed to bring about dramatic reductions in road fatalities. We turn now to another factor that forms a key part of all road safety strategies: engineering.

ENGINEERING

Changing driver behaviour is only part of the solution: we also need high-quality, well-designed roads. Road engineering is an area where national road authorities work closely with road safety bodies, prioritising roads for development and maintaining vehicle standards. We may have a long way to go in improving the quality of some Irish roads but Ireland has made considerable progress in elevating road standards in recent years.

Some people would suggest that improving road quality is an important first step, but that actually changing our mindset around road design and driving will lead to high-impact and long-term effects. One proponent of the rejection of traditional approaches to road design in favour of new and innovative perspectives was Hans Monderman. Monderman, a Dutch road traffic engineer, radically revolutionised road design and engineering. He proposed the concept of a 'shared space', where road markings, signs and traffic lights are stripped away and there is no distinction between footpath and road. Intuitively, this seems like a step backwards that would lead to chaos. With no delineations or markings on the road, how would a driver know when to slow down? How would a pedestrian know when to cross the road or a cyclist know who has the right of way? Surprisingly, far from the anarchy one might expect, Monderman's approach was considered highly successful. Sarah Lyall, writing in *The New York Times*, observed examples of his road designs first-hand and noted that 'in spite of the apparently anarchical layout, the traffic, a steady stream of trucks, cars, buses, motorcycles, bicycles and pedestrians, moved along fluently and easily, as if directed by an invisible conductor'.[1]

The principle behind this is simple. Road signs, lines, markings and rules (according to Monderman, at least) increase danger to road users by drawing attention away from what they should be looking at: other road users. When roads are redesigned with the aim of encouraging people to negotiate their movements on the basis of other road users, roads become safer and simpler to use. Lyall quotes Monderman's explanation: 'All those signs are saying to cars, "This is your space, and we have organised your behaviour so that as long as you behave this way, nothing can happen to you." That is the wrong story.'[2]

Monderman is not alone in this vision. A US-based organisation called National Complete Streets Coalition also aims to change the design and construction of roads, with an emphasis on all types of road users. While this specific model may not work in all contexts and is not without its limitations, it is this type of modern, exciting revolutionary thinking that represents the future for engineering research and design.

One crucial aspect of any such approach is a tangible shift away from the 'car-as-king' culture that has dominated in Ireland for the last number of years. Reducing the emphasis on cars is not only important for environmental reasons, but has been shown to have a demonstrably positive impact on road traffic collision data. If we look at countries like the Netherlands, where bicycles take priority, we tend to find some of the lowest road traffic fatality rates in the world. As Elizabeth Press, a filmmaker for Streetfilms, writes: 'The Netherlands is widely recognised for having the highest cycling rates in the world...They [cycle] because after the country started down the path toward car dependence, they made a conscious decision to change course.'[3] For environmental, safety and financial reasons, we may see a significantly less car-oriented approach to transport in Ireland over the coming years.

Overall, by challenging traditional philosophies and perspectives on driving, I envisage a future of progress and innovation in applied settings. The future of road safety, in terms of engineering, may lie in rethinking road design, broadening our perspective and being open to new possibilities.

Despite this, some people would suggest that drivers will continue to act recklessly as long as they are not getting caught. This argument has led to the inclusion of, and dependence on, enforcement in all road safety strategies.

ENFORCEMENT

'The best car safety device is a rear-view mirror with a cop in it.' This safety slogan accurately sums up the perspective that enforcing traffic laws is the one sustainable counter-measure against risky and reckless driving behaviour.

Enforcement relates to the application of a range of penalties to road users who break traffic laws. Such penalties can be financial, points-related or involve vehicle or licence withdrawal. In high-income countries, enforcement is increasingly becoming reliant on technology. The introduction of speed cameras, for example, has acted as a form of both deterrence and punishment. This type of strategy is likely to become increasingly common in the future, particularly given the disproportionate ratio of road users to traffic law enforcement personnel.

Sometimes (particularly in lower-income countries), however, the traffic laws or enforcement strategies are not appropriate, up-to-date or relevant to all road users. For example, some researchers would suggest that a number of road traffic collisions can be attributed to inappropriate speed limits. Historically, speed limits set on both urban and rural roads were based on the number of cars and other vehicles on the roads at that time. In the UK, the 30-mph urban speed limit dates back to 1934, when there were fewer than 2 million motor vehicles; today, that number has risen to well over 33 million. Researchers suggest that not all traffic laws, and certainly not all speed limits, have been updated to reflect the increasing number of road users.

The positive side to this is that ordinary people are stepping up to help combat these problems. Strategies such as the 20's Plenty For Us campaign have become widespread in Europe. This UK-based campaign, aiming to make 20mph the default speed limit in urban areas, has been ongoing since 2007. Since its founding, a large number of UK authorities (affecting over 7 million people in total) have implemented or are committed to implementing a 20mph default limit in residential areas. The success of campaigns like this is evident in the statistics. By the second year of the 20's Plenty campaign, the city of Portsmouth had reported 22 per cent fewer casualties. It seems that, where enforcement falls short, we are starting to take the initiative ourselves – and that, surely, is a cause for optimism.

ROAD SAFETY ON AN INTERNATIONAL SCALE

While road traffic collision data in high-income countries indicates a reduction in collisions, a recent World Health Organisation report has estimated that by 2020 road traffic collisions will be the third-leading cause of death globally. The reason for this is that the distribution of fatalities resulting from road traffic collisions varies enormously across developed and developing nations. A 2009 global status report on road safety indicated that more than 90 per cent of fatalities from all road traffic collisions occur in low-income and middle-income countries.

The good news is that this trend is not going unnoticed. Working professionals in road safety are aiming to reduce the discrepancy between low- and high-income nations. In international road safety conferences, for example, symposia dedicated to road safety in developing countries are becoming commonplace, and an increasing amount of research specific to developing countries is being conducted. The future for global road safety is in the development and application of evidence-based, context-specific solutions – a major challenge for researchers and policy-makers.

THE NEED FOR EVIDENCE AND RESEARCH

The need for evidence-based strategies has arisen repeatedly in this essay. In order to develop and implement these types of strategies, we need to conduct systematic empirical research. This relates to a fourth, lesser-known *E*: evaluation. The Road Safety Authority in Ireland, and road safety bodies elsewhere, currently provide a number of bursaries for research in this area. But the research area is wide-ranging: too often, research streams will work independently of one another, leading to a lack of over-arching theories and models. A multidisciplinary, interactive and integrative approach to road safety is where I believe road safety research and practice should be headed in the future. The future of road safety lies in an approach that incorporates theoretically-grounded education campaigns for young people, evidence-based enforcement strategies and more modern perspectives on road design.

The research I have done over the last three years has been, at times, disillusioning and frustrating. But I have never doubted that the type of research I'm doing is worthwhile, necessary and important, and I have never doubted that without road safety research, Ireland would never have made the kind of progress it has in reducing the number of deaths on our roads.

Research will inform our decisions, our laws and our policies – but there are some aspects of the human condition that are difficult to change. As humans, we have an 'optimism bias', leading us to believe that nothing bad will happen to us. We also have a 'self-enhancement bias', leading us to believe that we are better at things than other people. Finally, for the past few decades, we have been living in a culture that applauds and promotes 'getting there faster'. When these three factors are combined, it means we systematically overestimate our driving skills, underestimate our risk of being involved in a collision and are constantly reinforced for making it quickly from one place to another. This may be where the real problem lies.

Producing sustainable and realistic visions for the future involves changing our mindset around driving. What is now needed is an understanding and acceptance that faster driving is not the answer, that safety is more important than time saved, and that our lives are worth more than reaching our destination five minutes earlier. The future for road safety is bright, but the onus is on us to create positive and permanent change; change that will identify Ireland as a model to look up to, and change that will remain for generations to come.

Notes

1 Lyall, S. (2005), 'A Path to Road Safety With No Signposts', *The New York Times*, 22 January.

2 Ibid.

3 Press, E. <www.streetfilms.org>.

Chapter 18

Anthea Lacchia

Science

Anthea is a PhD student in Trinity College Dublin, undertaking research on goniatite fossils found in the rocks in the west of Ireland. Born in Italy, she moved to Dublin in 2007 and graduated from Trinity in 2011 with a first class honours degree in Geology and a gold medal from the college. She was also highly commended in the Sustainability category of The Undergraduate Awards of that year. A freelance science writer, Anthea was also Science Editor of Trinity News. *Her work now includes contributions to several publications, such as* Science Spin *magazine.*

The construction of a new bridge over the Liffey, which is currently underway in Dublin, has sparked the collective imagination of many people who, ever since its announcement, have felt compelled to come up with strange and often semi-serious names for it. Suggestions have varied from Brian O'Driscoll Bridge, to Mary Robinson Bridge, to Bailout Bridge, to Ernest Walton Bridge.

The latter was put forward by the Institute of Physics in Ireland to honour Nobel physicist Ernest Walton, Ireland's only Nobel Prize winner in the sciences. In 1932, he and his colleague John Cockcroft were the first to artificially split the atom in Cambridge, thus earning them the Nobel Prize in Physics. Given that Ireland is often remembered for its playwrights, poets, artists and musicians, but less often for its scientists, this could be a fitting choice.

From Jocelyn Bell Burnell, who in 1967 discovered pulsars (rapidly rotating neutron stars), to Robert Boyle, often called the 'Father of Chemistry', to John Tyndall, one of the greatest scientists of the nineteenth century, Ireland has been a fertile breeding ground for excellent science for many years – but the work and findings of Irish scientists often remains obscure. Is the subject matter too complex? Or are scientists not good enough at communicating their research? Is there a lack of interest on the part of the general public when it comes to science? Or is the so-called 'anti-science trend' alive and well?

In the eighteenth century, William Blake, who was opposed to the rise of Newton's scientific 'materialism', wrote: 'Art is the Tree of Life. Science is the Tree of Death.' Similarly, the Romantic poet John Keats accused Newton of having destroyed the poetry of rainbows by explaining the origin of their prismatic colours. However, understanding something hardly makes it less beautiful or awe-inspiring.

Even though this kind of open hostility is uncommon today, science is seldom considered a true component of social culture. In fact, sometimes it is unclear where to place science in modern culture and in our lives in general. Charles Percy Snow's 'The Two Cultures' essay (1959) laments the cultural divide between science and the arts.

> *I believe the intellectual life of the whole of western society is increasingly being split into two polar groups... Literary intellectuals at one pole – at the other scientists, and as the most representative, the physical scientists. Between the two a gulf of mutual incomprehension – sometimes (particularly among the young) hostility and dislike, but most of all lack of understanding... A good many times I have been present at gatherings of people who, by the standards of the traditional culture, are thought highly educated and who have with considerable gusto been expressing their incredulity at the illiteracy of scientists. Once or twice I have been provoked and have asked the company how many of them could describe the Second Law of Thermodynamics. The response was cold: it was also negative. Yet I was asking something which is the scientific equivalent of:* Have you read a work of Shakespeare's?

Since Snow's essay, much has been done to bridge the art–science divide and scientists in particular have become more aware of the need to communicate their findings. In 1985, Britain's Royal Society published a report on the public understanding of science (PUS), usually referred to as the Bodmer Report (after Sir Walter Bodmer, who chaired the working group that produced it). The aim of the report was to investigate whether or not the public was apathetic towards science and, if this apathy existed, what might be done about it.

Since then, there has been a move towards a two-way dialogue between scientists and the rest of society, and the emphasis has switched to public engagement with science. This two-way approach moves away from the idea of white-coated scientists emerging from dimly-lit labs with paper scrolls full of 'truths' ready to be divulged to the people. Rather, it seeks to gauge the thoughts of the public in reference to the possible future paths that technology and science can take.

After all, from black holes to targeted cancer therapies, from a raindrop to a supernova, we live in a wonderfully complex world. People are interested in understanding the workings of our universe – and science is one of the many avenues through which we can gain this understanding.

Science can engage with society in many ways: science sections in newspapers, blogs, podcasts, videos, social media accounts, science museums and popular science books. Not all of these are the domain of scientists, who at best can explore them in their spare time, but they are the domain of 'science communicators', a term that is somewhat ambiguous but designates people who take on a mediating role between science and society; their input is key if public engagement of science is to take place.

Ireland is home to many excellent scientists and science communicators and, especially since Dublin hosted the European Science Open Forum (ESOF) in 2012, issues surrounding the public engagement of science have been the subject of discussion. The conference was accompanied by a series of events all over Ireland which were aimed at the general public. For instance, Brian Greene, physicist and science writer, brought his science fiction reimagining of the Icarus myth, *Icarus at the Edge of Time*, to the Dublin National Concert Hall, with music by Philip Glass and images from the Hubble Space Telescope. In this futuristic tale inspired by a classical myth, Icarus travels not to the sun but to a black hole.

This is an example of art and science magically fused together to inspire and inform the public, children in particular, on the wonders of the cosmos. I see this kind of performance as a very effective means of connecting people to science and we can look forward to more of the same in Ireland's future. In fact, since July 2013, an annual science event entitled Festival of Curiosity (the legacy of ESOF) has been established.

In Ireland and on a global level, it is becoming more and more common for graduate science students to be coached in science communication. In addition, scientists at higher levels have access to media training courses, where they can learn how to write press releases or how to explain their research during an interview.

In February 2013, the Irish government announced a €300 million funding package for research in several areas of science. Significantly, Science Foundation Ireland (SFI), in charge of distributing this funding, sees public engagement as one of the four primary objectives in its strategic plan.

Because much scientific research is funded by taxpayers, there is general acceptance that the public should be informed of the research outputs. Having said this, most scientific papers are costly to acquire and, in any case, are incomprehensible to anyone outside a tiny circle of specialised workers.

But why should it matter in the first place? There is a growing consensus that citizens need to be *au fait* with leading scientific themes in order to make informed decisions on issues such as climate change, energy and health policies. This kind of evidence-based knowledge can inform the future of science.

My own field of geology is often the subject of interest in the public domain, the most popular areas tending to be fossil fuel extraction and earthquakes. The Greek historian Herodotus in the fifth century was perhaps one of the first to bring geology into the 'popular science' realm: in fact, in his *Histories* he describes how Egypt had been created by silt deposited by the Nile. Since then, countless popular science books have been written on the subject, such as *Earth* by Richard Fortey. In Ireland's geology community, blogging is slowly proliferating among postgraduate students and researchers. Geology-themed walking tours and university open days during which the public can learn about planet Earth through the examination of rocks and fossils provide the opportunity for earth scientists to share their enthusiasm with the wider public.

However, in geology (as in any other science) not all scientists are able to communicate their findings in an audience-friendly way – and I believe they don't all have to! Take Newton, for example: he was notoriously difficult to understand outside mathematicians' circles and even wrote that this was deliberate in order 'to avoid being baited by little Smatterers in Mathematicks'. Some brilliant scientists will simply prefer spending precious hours working on their latest experiment than thinking about how to translate their ideas for the general public, and so be it. These scientists are increasing our knowledge of the world, while doing what inspires them. This is where science communicators and science journalists come into play: they are the messengers, the translators.

As Stephen White wrote in *Successful Science Communication*: 'I want scientists to do science and journalists to do journalism – the essential bit is in the middle where the two professions meet each other for mutual benefit.'

It seems to be a fairly simple process: scientists want to spread their ideas and journalists are always looking for new stories to share with the rest of the world. Win–win, right? Unfortunately, science and journalism have not always had a peaceful relationship with each other. It's easy to see where tensions might arise. After all, a scientist may have been working on a particular branch of research for decades before finally making a breakthrough and announcing it to the world. It can be heart-breaking for a scientist if the results of several years' work are not reported upon accurately. It is a sad truth but many scientists have a lot of fear and mistrust when it comes to dealing with the media.

Of course, any good scientist and any good journalist would ask the same question: How do such inaccuracies arise in the first place? The truth is that journalists work in increasingly pressurised environments. A newspaper may want to run a story about the latest scientific discovery, but the time given to the journalist to actually research the topic, consult the relevant people and write up the article may be just a few hours! In the worst case, this can result in the hasty rewriting of press releases; otherwise, it can result in inaccuracies. Furthermore, when complex research papers need to be translated into language that is non-technical, the associated risk is a dumbing down or trivialisation of science. Often, in the rush to simplify, speculative remarks about the importance of a given result are reported as actual findings. This goes hand in hand with a tendency to sensationalise science, to make the story 'sexy', to report the latest 'major breakthrough'. But science isn't all about major breakthroughs: it's often about the latest tiny incremental step towards a new discovery.

Scientists also complain that only certain types of stories get picked up by the media, usually the ones with the most applications. The truth is that journalists want stories that will engage and interest their readers. They need to capture the reader and draw them in. In order to do so, the first sentence is key: this is usually the point at which the reader will decide whether or not another section of the paper merits more attention.

There is a debate as to whether journalists with no scientific background should be writing about science in the first place. You may be an accomplished news journalist with a knack for breaking the story, but hazy secondary school memories from chemistry and physics classes might not be enough to write about the latest developments in nanotechnology. In my opinion, science journalism requires a minimum of scientific training – and senior editorial staff need to take this into account. With these issues in mind, I feel sure that we can avoid the grim situation portrayed by Ben Goldacre in *Bad Science*: 'the media create a parody of science… Science is portrayed as groundless, incomprehensible, didactic truth statements from scientists.'

In Ireland, the links between scientists and the media are actually pretty well established, with many scientists knowing how to best spread the word about their research. I see the future as building on the existing mutual respect and trust between them. In the context of universities, research groups should have a clear communication strategy and co-operate with communication professionals.

Even as we move into a world that is increasingly fast-paced and where scientists are required to put on the communicator's cloak and journalists are required to put on the lab coat, all while tweeting and writing up three stories in the space of an afternoon, there will always be room for the science communicators – those people with the precious ability to explain, excite, educate and simplify. These people need to be able to hop across borders of different disciplines, within science and also between science and the arts.

Breaking down the barriers between disciplines and increasing interconnectedness between them is something that has been championed by the Science Gallery, an initiative of Trinity College Dublin. In truth, one can't talk about public engagement of science in Ireland without mentioning it: the Science Gallery's exhibitions explore the intersection of art and science, looking at the ideas, discussions and feelings that are generated when people are brought closer to both worlds.

In fact, great ideas often stem from interdisciplinary approaches but, in the words of David Edwards (founder of Le Laboratoire in Paris, another creative art–science centre): 'Those who cross artscience territory sometimes experience loneliness, institutional discouragement, and even fear; but having overcome the resistance and explored this novel territory between the arts and sciences, they often find it so much to their liking that they never leave. They become artscientists.'

One art piece that was featured in the gallery was called *Victimless Leather* and it consisted of a little coat made of animal cells growing on a matrix. When confronted with it, the visitor was forced to reconsider what 'victimless' means. No animal was killed but the cells are still animal-derived, so is the coat really victimless? This leads us to further consider our ethical responsibility as consumers and our perception of biological life. As designer and curator Oron Catts explains, the piece 'highlights the divide between cultural perceptions of life, which are often incompatible with the study of life in scientific laboratories'.

Whether it is in abstract painting, poetry or literature – for instance, think of Mary Shelley's *Frankenstein* and science fiction in general – it is easy to see that science can inspire art. But likewise, art can inform science and the importance of creative moments in science is often underestimated.

Ireland's science-themed future is bright: it is a place where we can not only enjoy the benefits of science and marvel at its applications, but also have a say in the future paths research can take. It's up to us: let's open up the boxes and increase the connections. The next time you're at dinner, why not surprise everyone with your knowledge about life on Mars or the Higgs boson? Let's put science on Ireland's menus: *I'll start with some nanotechnology, with a side of palaeontology and finish off with some renewable energy for dessert!*

As we move forward, I envisage a beautiful bridge between science and society, built on open-mindedness, dialogue, art, music, performance, panel events, conferences, social media and the web. By building this metaphorical bridge, Ireland can play a leading role in science communication in the global context and in merging the two cultures. Who knows what surprises will come out of this union?

Chapter 19

Darren Ryan

Social Entrepreneurship

Darren is Head of Engagement at Social Entrepreneurs Ireland, a fantastic organisation that funds and supports social entrepreneurs, helping them to increase their impact. Having graduated with a BA in Economics and Politics from Trinity College Dublin, Darren first spent time working in overseas development in South Africa and Kenya, before turning his attentions to social change closer to home. An Ireland Funds young leader and a member of Sandbox, he is also co-founder of the Suas Alumni Network. More recently, he founded Snooze Academy, a fun initiative that encourages people to defeat the snooze button and reclaim their mornings.

The stage is set for Ireland to lead the way globally in our approach to tackling societal issues. For too long have we looked jealously to the Scandinavian countries with their excellent records of social development. We have everything we need right here in order to prosper: a nation that is built around community; a talented pool of entrepreneurs; a highly-skilled and educated population; and the good will of a dedicated diaspora that stretches around the world. And now, following the events of the last few years, we also have a strong desire for change, a demand for new approaches and a renewed commitment to developing the country.

With these building blocks, we have a unique opportunity to change the way we deal with social problems – and this new approach could set us apart from the rest of the world. But, although we have many of the necessary conditions for progress in the social sector, there are also a number of significant obstacles. As with any system, there is a tendency towards inertia and, despite our renewed enthusiasm for change, there remains a reluctance to actually challenge existing models and ways of working. We are also confronted with a fast-moving world that constantly throws up new challenges. Furthermore, the current recession leaves us with very limited resources available for social investment. In many aspects of society, we are running just to stand still.

In order to overcome these obstacles, we need to find and harness new approaches to tackling our social problems and we need to develop flexibility in delivery of services. We need to focus on addressing the root causes of society's most pressing problems, rather than battling against the symptoms. And we need to enlist people who make things happen – visionary leaders who will be driven to bringing about fundamental change in our society. We need social entrepreneurs.

If Ireland is to fulfil its potential as the global leader in the social sector, what role can social entrepreneurship play? For that matter, what exactly is social entrepreneurship and how is it relevant to Ireland?

While social entrepreneurship can be defined in many ways, it is fundamentally about developing innovative solutions to social challenges. In areas as diverse as healthcare, education, social care, mental health, migration, employment, unemployment and disability, the opportunity exists to rethink our approach, to investigate the latest methods from around the world and to take the necessary actions in implementing them. Just like commercial entrepreneurs, social entrepreneurs create value. However, in the case of social entrepreneurs, the value they create can be measured in social impact rather than financial profit.

As we look to the future, social entrepreneurship must play three key roles in bringing about change in Ireland. Social entrepreneurs must be the catalysts for change. Social entrepreneurs must also be the first responders to change. And, as a society, we must ensure that we unleash the power inherent in many different sectors.

CATALYSTS

Social entrepreneurs will be the catalysts that turn the desire for change into practical solutions and actions. To quote Andy Warhol, 'They say time changes things, but actually you have to change them yourself.'

Even in the 'good times' in Ireland, we struggled to make a significant difference to some of our entrenched social problems. As we look to our ideal future, it is easy to assume that some of the challenges we face today will be fixed over time; that the problems are transient.

However, the truth is that our society as it stands, with all its unmet needs, is likely to remain stagnant at best. With falling resources in the social sector and a rise in demand for services, many organisations are struggling just to keep services at their current levels.

Social entrepreneurs will play a vital part in the progression of change, acting as innovators who test and adopt the best new approaches to solving social problems. We need social entrepreneurs to provide different approaches, to find access to previously untapped resources and to take the action required to bring about change.

FIRST RESPONDERS

The other side of social entrepreneurship is the ability to respond quickly to developments in the external environment. The world around us is certainly in flux. While we're already struggling to tackle entrenched social problems, the challenges that we will face in the future cannot even be predicted.

On the other hand, emerging technologies and new approaches that have worked elsewhere provide wonderful opportunities to transform how we tackle social problems. While we know that new opportunities will emerge, we don't yet know what they will look like or what their potential impact will be.

Although we cannot predict what the problems of the future will be, we know that we will need a rapid response if we are to address them properly. In the same way that social entrepreneurs have emerged in recent years to tackle issues surrounding unemployment, or reacted swiftly to apply technological solutions to other social problems, we will also need timely responses to future developments.

While existing organisations and the public sector can be slower to respond to new approaches (especially with the restrictions brought about by size and structure), social entrepreneurs can provide a fast and effective response. They are not bound by current practices, nor are they slowed down by bureaucracy. They may also have a unique understanding of the issues that are emerging; a social entrepreneur's commitment to tackling an issue often derives from their personal experiences.

UNLEASHING THE POWER OF DIFFERENT SECTORS

The challenges that we will face in Ireland in the future will be too complex to be solved by any one sector or any one kind of organisation. While each of the established organisational structures will remain vital to a flourishing economy and society, there are some limits to each of them that will prevent maximum social impact. For example, charities may have the benefit of being able to secure donations and draw on voluntary labour, but they may lack the ability to raise investment capital and they may find it difficult to embrace risk. Private companies can draw down investment and loans and generate income, but they may have responsibilities to shareholders and investors that compete directly against creating a social impact. The public sector has the benefit of scale, but it often moves slowly. Each of these sectors has an important role to play but when they are acting in isolation, the weaknesses of each sector prevent wide-scale social innovation from flourishing. It is only by breaking down the barriers between these sectors and taking the best from each that we can hope to solve problems of the scale that we are sure to face.

Social entrepreneurs can draw on the best elements of each traditional sector, blurring the lines between them.

We can move towards a future where all organisations are mission-driven, focusing on *impact* as much as profit as the core driver of activities. This alone would be a massive shift in the way we view the different sectors, smashing the false dichotomy that has emerged between commercial and social entrepreneurship. Mission-driven organisations led by social entrepreneurs will bring the best of both sectors into play. Entrepreneurs of the future will first decide the impact they want to have and then choose whatever organisational structure best gets them there, whether that is a charity, a corporation or a hybrid model.

A BLUEPRINT FOR SOCIAL ENTREPRENEURSHIP

There are great benefits to fostering a culture of social entrepreneurship in Ireland, and thankfully the movement is gaining considerable momentum and widespread interest. Social entrepreneurs are inspiring and they are particularly effective at bringing people along with them. As well as coming up with great ideas, they take action to bring about change. Ireland's long history of strong community involvement, coupled with our impressive cadre of entrepreneurs, makes us an ideal location for embedding social entrepreneurship.

However, the growth and success of social entrepreneurship in Ireland is far from assured. While the momentum behind it is certainly considerable at the moment, in order to make a lasting and positive impact, we will need to create the right environment in order for it to flourish. This will require co-operation and innovative thinking from all sectors of Irish society.

As the largest spender in the social sector, government needs to create the space for innovation to take place. For many reasons, government is not designed to innovate. However, it can be the driver that ensures innovation is taking place and, even more important, it can also remove itself from certain roles in order to allow space for innovation.

While government should be expected to set the level of services that are required in a society, it does not necessarily need to deliver these services. Any large organisation dealing with human-centred services (e.g. health or social care) will struggle to provide adaptable solutions that are appropriate for a wide range of people, all with differing needs. The public sector could unleash a wave of innovation by allowing flexibility in service delivery, provided that key outcomes are achieved. Rather than government dictating the precise nature of such delivery, contracts could be put to tender based on desired results, allowing social entrepreneurs to innovate, test new approaches and deliver services more effectively and efficiently.

In order to fast-track social entrepreneurship as the key driver in the Irish social sector, we will need to create a small number of hugely successful case studies. In the short-run, this means investing heavily in the social entrepreneurs with the most potential to create social change. Doing so will show that, when provided with the right tools and support, social entrepreneurs can transform the way a problem is dealt with. This will require funders (both public and private) to open themselves to more risk in their social investments and to increase the level of funds flowing towards social entrepreneurs. Ultimately, the level of investment support will need to be well in excess of what is currently provided in the sector.

Social entrepreneurship should act as a complement to, rather than a replacement for, existing networks of community and voluntary agencies throughout the country. Ireland has a fine tradition of community involvement and delivering local responses to social challenges. Social entrepreneurs will need the support of existing organisations in order to replicate, scale and franchise the most effective social solutions so that they can reach as many people as possible, as quickly as possible.

Existing non-profit organisations can also play their part in the process of innovation, by being open to challenging and rethinking their current structures and approaches and by constantly looking for ways to improve the services that they offer.

Currently, there are limited options in terms of company structures available to entrepreneurs who want to play in the grey areas between charity and corporation. Social entrepreneurs who are thinking big and who want to fundamentally change things in Ireland can often be frustrated by the lack of funding available. Yet Ireland has a flourishing private investment market and excellent networks of private investors. Finding a way to open up private capital to leading social entrepreneurs will be a vital step towards transforming the scale at which we tackle social problems. Already a small number of socially-focused entrepreneurs have chosen to set up as private companies in order to avail of private funding, offering equity as a method of raising the money they need to increase their impact.

A legally recognised hybrid company structure would enable socially-driven organisations to maximise the resources available to them by offering something to both private investors and donors. This new structure could also be used by traditional businesses that recognise that long-term social sustainability is vital for long-term financial sustainability. Dismantling the existing either/or structure will open up opportunities for socially-focused organisations to become more business-based and for businesses to develop a stronger social focus.

Imagine if the most entrepreneurial minds in Ireland were the ones charged with solving society's biggest challenges? Now is the time to reclaim entrepreneurship from a narrow business perspective and establish it as a broad approach to creating value, whether that's financial, social or cultural.

> *Entrepreneurship is the pursuit of opportunity without regard to resources currently controlled.*
>
> Professor Howard Stevenson
> Harvard Business School

As technology continues to develop, the potential to apply technological solutions to social problems will certainly grow. This will yield exciting results. Why would a young entrepreneur want to spend their time building a run-of-the-mill app when they could be utilising the same technology to change somebody's life – or indeed even save a life? By directing our many highly-skilled tech entrepreneurs towards tackling our social problems, we can unleash a wave of fresh ideas, passion and action to this area.

> *It is not the strongest of the species that survives, nor the most intelligent that survives. It is the one that is the most adaptable to change.*
>
> Charles Darwin

As we move into unknown territory and the world around us changes ever faster, so having a social sector that is flexible, fast-moving, open, responsive and innovative will be crucial.

Mobilising solutions to current problems, anticipating and responding quickly to future problems, combining sectors and tackling the root causes of problems will be the core benefits to Ireland of embracing and supporting the development of social entrepreneurs. How we get there will be dependent on many external circumstances, but we can pave the way for the emergence of social entrepreneurship by being open to innovation in delivery of public services, building inspirational examples of success, mainstreaming the best ideas, creating innovative business models and attracting the best talent to focus on social challenges.

We can create a society of which we are proud, we can inspire other nations to do the same and Ireland can become the world leader of social change.

Chapter 20

Matthew Smyth

Theatre

Matthew is a theatre producer, writer and comedian based in Dublin. A Rough Magic Theatre Company SEED*, he has worked on acclaimed productions such as* Tender Napalm, FLATPACK *and* Heroine for Breakfast *and is one-quarter of the comedy troupe A Betrayal of Penguins. A member of Sandbox, he is also co-founder of Collapsing Horse Theatre Company, which has produced* Bears in Space *and the gorgeous puppet musical* Monster/Clock. *He co-ordinates the annual summit of The Undergraduate Awards.*

As I started typing this, Fulham FC banged a goal past my beloved Newcastle and the Magpies slumped to what is yet another defeat this season. I was 5 years old when I decided to back the wrong horse and follow Newcastle United, despite everyone in my class choosing to cheer for Man United or Liverpool. Those classmates have all since enjoyed their share of the glory years, while I spend most Saturday afternoons praying for a draw.

I was 22 when I decided to back yet another floundering foal, this time a career choice: Dublin and its theatre scene. Or at least I was told it was a bad bet. Like in junior infants, I ignored all the pros of picking the other more successful teams and stuck with my gut feeling. What's more, as friends and loved ones all involved in the arts moved across the pond to the more expansive scene, I hung around Dublin. Granted, most of that particular decision was down to a combination of laziness and fear of the unknown.

Thankfully, it worked out in my favour. Because of the good grace of Rough Magic Theatre Company, when I was three productions into my burgeoning career I was taken on as a SEED. The SEEDS programme is a training programme that seeks out, enables, encourages and develops budding theatre talent. Along with the actual mentorship programme, the prestige of this label from Rough Magic meant that I was sought-after by every young and middle-of-the-road theatre, opera and dance company on the go. To say I'm grateful is a very serious understatement.

So, without putting a massive amount of forward thinking and planning into it, I became a producer on the Dublin scene. A few noteworthy productions later, mainly as a result of surrounding myself with all the talented people I know, I've inexplicably become a 'successful' producer. This would certainly have made it easier and much more feasible to go pond-hopping, but again, my gut has kept me here. Although, I'm sure laziness is still playing its part in that regard.

Conversely, the biggest thing I learned from the first year of my mentorship is that Dublin is not at all the den of inequity I had once been told. It is packed full of talent, opportunity and people who care so much about what they do. Of course there are Buts: but people do still leave; but companies still shut down every day; but modern Irish theatre, internationally, is still seen as some weird cousin of the vibrant English theatre scene. Yet here I remain, and here (I'm happy to report) I have managed to earn something of a living.

Having said that, the biggest But of all is this: but there's no money. Arts Council funding is shrinking rapidly and, by their own admission (and despair), will continue to do so. However, most artists still find themselves in the funding loop: by which I mean continuing to apply with the hope that if you throw enough things at the wall enough times something will eventually stick.

So the problems of Irish theatre are plain to see – even for a rookie like me.

> *That's the problem with planning a late-night supper after the opera: not only does the hero or the heroine die singing, but you end up famished after the last notes of the finale.*
>
> E.A. Bucchianeri, *Brushstrokes of a Gadfly*

Money. Of course it's always about money. No matter how dedicated and engrossed you are in your art form. No matter how prepared you are to live the life *La Bohème*, it's always going to come down to money in some shape or form. (Well, the shape is usually rectangular and it's usually in note form.)

Times have been tough in the past but it has never stopped good things happening in theatre. Galway's Druid Theatre Company erupted from the 1970s against all odds and it is now renowned across the globe. Rough Magic is the prime example of a 1980s success story. So let's not claim that a downturn is an obstacle to the arts: more often than not, the opposite is true.

With quality as a base, these companies attacked the market in their own new and unique style. So with what niche will the next generation go and look for an audience, without any marketing budget (apart from social media manipulation, of course)? Companies both old and new are always experimenting to recapture the imagination of theatregoers, but at what point will the funding model adapt to the realities of today?

Perhaps we should first ask: why fund the arts at all? When times are tough, perhaps we should be channelling our limited supplies elsewhere. This surely must seem obvious to the likes of those running the country. They know that some day we may need a hospital rather than, say, an experimental version of *Hamlet* performed by rare red squirrels. But they would be forgetting that the arts are a necessity also. When you finish work, what do you come home to? Television, books, a painting in your house, your CD collection – all of these things are brought to you by 'the artist'.

When Winston Churchill was asked why he wasn't diverting money from the arts into the war effort he answered with another question: 'Then what are we fighting for?'

But perhaps there is a better model. Cash-strapped startups in the entrepreneurial world, for example, manage to raise money from corporate venture capital and other sources of funding that are far from government-based. While there is an element of corporate sponsorship in operation in the Irish arts today, and it is true that philanthropists still prop up certain artistic endeavors in this country (although this happens on a much smaller scale than in the US, for example), one has to wonder if the arts couldn't give these tech startups a run for their money by adopting some form of venture capital model. People in the arts certainly have all the creativity required.

I genuinely believe that the future of sustainability in the arts lies in a new approach to funding. The Arts Council is already recommending that arts companies take on a fundraising role, although the problem is that the Arts Council is giving this advice to companies whose staff are already stretched to the limit.

Let's put the power back in the hands of the audience then. At the moment, the Arts Council has a very large say in what kinds of productions Irish audiences will see; this is because they have the power to fund the plays. Only after this funding has been secured does a performance go ahead, meaning ticket sales in these instances are a second source of funding. I believe prospective audiences should have more of a direct say in what plays they'd like to see. And an audience decides what they're going to see in the same way they have been doing for thousands of years: with their feet.

I've only had the pleasure of having an Arts Council-funded show once in my career so far. (*FLATPACK* by Ulysses Opera Theatre Company was an opera about IKEA. Yes – really.) It meant that, even if the ticket sales and the reviews had been terrible, everyone would still have been paid. (Thankfully, it did extremely well.) The funding reality is that such a safety net will be harder to secure. And, you know what? Maybe that's not a bad thing.

In this reality, the companies that survive will be the ones that don't just take what the audience wants courteously into account, but actively go and find out what performances the residents of a city or town actually want to see. This will see a shift in the feedback network, which is unfortunately missing from how theatres, companies and artists operate in Ireland at the moment. Even if companies have ample corporate or philanthropic backing and aren't depending on an audience, it's likely that they are still answerable to a set of requests from the financial backers.

The perfect vision is an environment that gets people excited about this art form, all the while keeping the artists happy and proud of the work they're producing. You don't want these artists to have their vision stifled. You also don't want them to be starved, or worse: to give up, like so many do.

In the midst of typing this essay, an interesting headline popped up on my newsfeed: 'Arts Council in UK to Face £11.6m of Further Cuts this Year'. Next thing I know, the final whistle is blown in the Newcastle vs Fulham game. The Magpies fall to yet *another* mediocre London team… It makes me think of the poorly-chosen-horse situation again. Maybe Newcastle really was the wrong team for me. Maybe theatre in Dublin was the wrong choice. Maybe theatre *itself* was the wrong horse to back.

But then I think about how hard people work in this industry. I can't attribute that to blind faith or loyalty – there must be something else there. There must be something worth fighting for; something worth evolving. The fact is that I would never want to leave this group of people who inspire me and everyone else around them.

This notion of evolving in order to meet the demands of the audience isn't an entirely new thing. There has been, and will continue to be, a shift of motivation for certain art forms. This can be seen even with new companies like WillFredd Theatre doing two shows in two years with very specific demographics in mind. One year their production *Follow* used sign language, which brought theatre to a whole new group of people who, for a long time, were entirely excluded. Theatre really is starting to speak to people in their own language. Already, there's a shift in how and why productions go ahead. It's a challenge, definitely, but it's also a huge opportunity.

All the world's a stage…

William Shakespeare, *As You Like It*

During my college years my peers and I spent every summer at the Edinburgh Fringe Festival, which became our primary location to see/research/drink to theatre each year in a veritable flood of international culture. As a result, the work that my contemporaries and I started creating on our return home was significantly different from everything else that was being made in Ireland at the time – there was no apparent context for it. Boy, did it stand out.

Whether it's for better or worse, I see one of the larger changes in arts and culture in Ireland being an increased international element. This might mean that work from abroad finds more of an audience here. It might mean that domestic productions find it easier to travel around Europe and further afield. Better still, it might mean both. Globalisation means I can easily follow the work of other international companies like Complicite in the UK or La Fura dels Baus from Catalonia, but the odds of them regularly coming to Ireland are as low as Newcastle's chances of staying up in the Premiership this year.

Another reality may be that as funding shrinks and companies start to fall off the map, we may have to bring international companies in. Extreme? Yes. Impossible? No. This will not only give future Irish practitioners a greater understanding of what's happening in other countries, but it will also give the audience an idea of what standards they can, and should, expect. Yes, we currently do have the Dublin Theatre Festival, the Fringe Festival and the odd touring production, but there's no constant, year-round source of theatre and culture from beyond our national borders.

It might seem counterintuitive to create competition for our domestic product in this way, but the long-term benefits could be huge. The idea is that the introduction of works from further afield would encourage and inspire domestic practitioners to pick up their game. It could also mean that the 'thesp diaspora' return to Ireland, if not to work with Irish companies then to at least perform with international ones on their home soil. What I would love to help build is a two-way conveyor belt between Ireland and the international scene, rather than the catapult-and-slingshot system we seem to have with our talent at the moment.

I want to burn with the spirit of the times. I want all servants of the stage to recognise their lofty destiny. I am disturbed at my comrades' failure to rise above narrow caste interests which are alien to the interests of society at large. Yes, the theatre can play an enormous part in the transformation of the whole of existence.

Vsevolod Meyerhold

In the UK, some theatre directors are household names. Runs of shows sell out before the cast even get into the rehearsal room. In Ireland, people often don't know who the flamboyantly-dressed 'creatives' are behind even the most important productions in their country. I would love to change this.

In London, there are so many fantastic social and theatre hubs accessible to people – whether they've bought a ticket to a show or not. The West End, the South Bank and central London are hives of nightlife and activity all based around theatre, or at least the wider cultural pursuits. Is it true to say that simply hanging out in these areas increases people's chances of buying a ticket to a show? Probably, yes. Does it happen in London? Definitely.

So why has no one attempted this on a grand scale in Dublin or elsewhere in Ireland? If the pub culture is so dominant in Ireland, why hasn't the arts scene tried to generate some extra revenue from this, if for no other reason than to create footfall into the venue?

I see this in practice in my own life. Every time I go for a coffee in the Irish Film Institute it makes me about 50 per cent more likely (although this is an unscientific assessment, I'll admit) to go to a film that evening or the next. As I drink my coffee, I'll always read the brochure on the table. And even if I don't go to a film that night, at the very least the IFI will have sold me a coffee. Since the IFI can make this effort, can other establishments go a bit further to encourage footfall?

I regard the theatre as the greatest of all art forms, the most immediate way in which a human being can share with another the sense of what it is to be a human being.

Oscar Wilde

None of the ideas in this essay can be taken by themselves to be the sole evolution required of theatre in Ireland over the next twenty or thirty years. If anything, this essay is a collection of musings and ambitions that I'd love to see realised because I genuinely think they can make a difference to the scene and to the country.

People in theatre don't always have a great craving for leaving a legacy. But there's something about being in Irish theatre at this moment. Everyone in the industry is very open about the issues and problems that have to be faced, so you can't help feeling like maybe you can be part of the change that's needed.

Maybe it's because I studied Economics in university, instead of English or Drama. Maybe it's because I'm entering this sector at an interesting time: when the way things were done is coming to an end without any real alternative, and the people who are living in the midst of this change can feel a bit…lost.

Any of these things would make you want to create something new – a new model, a new type of company, a new approach. It doesn't really matter whether we use the ideas in this essay to find our way. What's important is that we try *something.*

Irish theatre is worth the effort, so I'm sticking with it.

I'll probably stick with Newcastle as well. I have way too many jerseys at this point.

Chapter 21

Tara Duggan

Waste

Tara graduated as an environmental scientist from University College Cork in 2010 with a first class honours degree. She was awarded a Beaufort Marine Bursary from the Marine Institute, which allowed her to take part in aquaculture research in west Cork and Kerry. In the 2010 Undergraduate Awards, she was winner of the Agricultural and Veterinary Sciences category. In 2011, Tara began researching the effects of vermicompost (compost made by worms) and later helped to set up the Celtic Worm Company, which specialises in farm waste recovery.

In 2010, the average person in Ireland generated 621kg of waste. This quantity of waste generated by our population would fill one football stadium every day, and we are quickly running out of landfill space to contain it all. The latest Environmental Protection Agency report shows that Ireland landfilled 1.5 million tonnes of waste in 2010. At this rate, we have only twelve years left before all the existing landfill space is full.

The good news is that we have plenty of alternatives. We just need to make better use of these alternatives, be more conscious of the volume of waste we generate and more aware of what happens to it.

Three-quarters of municipal waste (waste generated by households and businesses) is biodegradable, of which nearly half is still landfilled. Ireland met the EU target for biodegradable waste recovery in 2010 but, according to the Environmental Protection Agency, we are still 'at risk' of not meeting the next target, which is due at the end of 2013. Of the 53 per cent of biodegradable waste that was recovered, composting was one of the main treatment methods.

Composting is the breakdown of organic material by micro-organisms in the presence of oxygen. In a composting pile there are billions of micro-organisms, bacteria and fungi, which break down the waste and turn it into compost. The action of breaking down the waste generates a lot of heat – up to 90°C – and this heat is controlled by composting facilities, which use it to sanitise the waste, killing off any unwanted pathogens and weed seeds. Composting is a natural process, harnessed by people to treat biodegradable waste, reduce its volume and turn it into a re-usable product: compost.

Another type of composting is vermicomposting. This is where the waste is broken down by bacteria, fungi and, in addition, worms. Tiger worms are the most common type of worm used for vermicomposting. Also known as red wigglers, they have very distinctive orange stripes and can eat up to their own body weight in a day; unlike earthworms, they are happy to live in large groups. Once the waste has been broken down by the worms, it is known as vermicompost or worm cast. Worm cast looks quite similar to compost, although it is more concentrated and has a finer texture. It can also contain more soil bacteria than compost, since it is produced at lower temperatures.

Composting municipal waste is difficult in Ireland, mainly due to the quality of the waste received by the composting facilities. This might sound a bit strange: it's waste, so it's not supposed to have 'quality'. This might be the case for waste that is being landfilled or incinerated (where it is no longer going to be used), but with composting and other waste recovery activities like glass and plastics recycling, the quality of the receiving waste is vital – especially when you want to turn something from a waste into a product that someone will use again. To be frank, if rubbish goes in, rubbish comes out.

Treating 'brown bin' household waste can be particularly difficult. For instance, brown bin wastes are often contaminated with plastics, metals, batteries, tools, household appliances and other objects. Anything and everything can end up in a bin that should only contain 100% biodegradable waste. With all these contaminants entering the composting process, the choices are to spend a small fortune trying to mechanically remove them (which is frequently unsuccessful) or to consign yourself to the fact that the compost you will be making is going to be awful and will have no use other than for the covering or capping of landfills. (At the end of each work day, landfills must use a cover material, the aim of which is to cover the fresh waste, prevent it from being blown around and reduce odour. Low-quality compost is commonly used to cover landfills.) Essentially, if you separate your food waste so that it can be recycled, much of this can still end up in landfill as landfill cover.

Plastic is by far the worst contaminant, mainly because of the large amount of it and the effect it has on the appearance of the compost at the end of the process; flaked throughout with multicolour pieces of plastic, not something anyone would want to use in their back garden or spread out on their fields. It is not the only contaminant though: glass and heavy metals are also an issue. If compost has too much of these contaminants, it can't be used, since it can be a hazard to the user and can also contribute to soil pollution.

What is the point in composting waste and diverting it from landfill, only to produce something of such a low standard that it has to be put back into landfill as landfill cover? Composting facilities are more than capable of producing – and more than willing to produce – a quality product that can be used in gardens, landscaping and agriculture; a product that works well to condition the soil and fertilise plants. After all, it is in their interests to produce compost with value. Unfortunately, low-quality incoming waste is holding the industry back.

These days, most people will have at least two waste collection bins, if not three or four. Separating our waste was pretty much an alien concept only a decade ago, but now it's like second nature; this is brilliant. However, a large volume of the segregated brown bin waste contains contaminants. My own house is no different: I often come home to find tea bags in the plastics recycling, which drives me bananas! It is an easy mistake to make, putting something into the wrong bin, but it does have a serious effect on the processing of that waste. Even putting a small amount of green glass into the white glass section of the bottle bank can ruin the whole batch, resulting in it being landfilled or used as a low-quality material (e.g. insulation), rather than being re-made into glass containers.

So can people be persuaded to be a bit more careful when segregating their waste? I think the right information is widely available. I know that on my waste collection bins there are large colourful stickers depicting what is allowed and not allowed in each bin. Maybe a closer association with the end product might help. For instance, people could visit their local composting facility and avail of free (or low-cost) compost. There are some sites that already give away or sell compost at a very low price to their customers but many can't, since the quality of their compost is too low.

I think the uptake of this type of system could be very high. Take food, for instance. More and more people are becoming aware of the quality of the food they are eating and many have even taken to growing their own, either in their own gardens or as part of an allotment scheme. By doing this, they can save money and guarantee the quality of the vegetables they eat. If people can be incentivised, by gaining access to low-cost or even free compost, then they might be more likely to separate their waste more efficiently, benefiting both the composting facility and the householder. I also think that the circular nature of this type of system is amazing. Think about it: a person sends their brown bin waste to be composted, collects the compost from the facility and uses it to produce food, the waste of which they send back to the composting facility, and so on. What this would produce is a really nice, closed system of waste management and recycling.

Penalties for putting the wrong waste in the wrong bin could also be employed. If the waste collector simply refuses to collect contaminated waste, then the householder would be required to be more careful. (Some contaminants are very easy to spot, such as plastics in the organics bin or vice versa.)

Personally, I don't think a penalty system would be as effective as incentives. In a survey that looked at Irish attitudes to waste management,[1] it was observed that people were unhappy with the amount of waste that was *sent* for recycling but wasn't actually recycled (largely due to contamination). When contamination is reduced, people will be happier knowing that when they go to the effort of separating a product, that product will actually be recycled.

Recently, Ireland launched a new compost quality standard, combined with a compost quality assurance scheme. These were launched to provide assurance to customers that the composts they are using have been independently verified. A new logo will now appear on the packaging of composts that meet these standards.

These standards were introduced because previous market research showed that a lack of independent verification created a barrier to the market. The logo means that the compost has been made properly, meets a strict quality standard and will be fit for its intended purpose. Hopefully, this assurance will generate more sales of compost in Ireland.

Compost has a wide variety of potential uses. It can be used to make potting compost or potting compost mixtures, as a top dressing for gardens and landscaping, as mulch, or even applied on a larger scale to agricultural fields and contaminated land. Compost supplies food for both plants and soil. For your plant, compost supplies the macronutrients (nitrogen, phosphorous and potassium) as well as the micronutrients (calcium, magnesium, copper, zinc, etc.) that feed the plant as it grows. As well as fertilising plants, compost also feeds your soil. In my opinion, a healthy soil is the key to a healthy plant. Compost contains a large amount of organic matter, which feeds the micro-organisms (bacteria and fungi) in the soil. Compost can also improve the physical properties of soil, such as soil structure, moisture retention and ability to retain nutrients.

Worm cast provides these benefits (and many others) to soil. As worm cast is generally more concentrated than compost, it is used in smaller amounts – making up around one-fifth of the soil. The nutrient content of worm cast usually contains more plant-available nutrients than traditional composts; and since worm cast is made at lower temperatures, the bacteria and fungi in it are closer to what you would naturally see in soil. The action of vermicomposting can improve compost quality, increasing plant growth and stress tolerance. The gut of a worm is also full of good bacteria: as the waste passes through a worm's gut it is coated in this bacteria, which is then transferred into the soil.

Both compost and worm cast can also have a bio-stimulant effect, whereby other beneficial properties of composts/worm casts (not just the nutrient effect) enhance plant growth, yield and vigour. This bio-stimulant effect can reduce nutrient input while retaining yield, and can also help plants to cope with different stresses and diseases. How these soil amendments work exactly to cause this effect is still the focus of much research worldwide, but it is an exciting area for development, especially now that prices for agricultural inputs (e.g. fertilisers) have gone sky high and finding alternative and sustainable replacements is an important step in reducing the cost of food production.

Compost needs to have value and, therefore, it needs to be of good quality. Proper segregation of waste and reducing contamination will go a long way towards achieving this. One waste that can often get misused is farm waste. Farmers use farm manure and slurries as part of their nutrient management plan, but they often have excess manure and slurry that they spread on their land unnecessarily if they have the capacity – or else pay for it to be taken away and treated elsewhere, which is very expensive. This excess manure is an issue because if it is spread where it is not needed, or at inappropriate times (e.g. during winter or before heavy rain), many of the nutrients are wasted and get washed into local water bodies. This, of course, contributes significantly to water pollution. If the farmer could deal with excess manure and slurry in a more sustainable way (i.e. composting) and get paid for this compost, the farmer benefits from an added income source and water pollution is reduced. Farmyard wastes are also more consistent than other waste streams. Usually only one or two people deal with the waste on a farm, so it is easier to reduce contamination on farms than it is elsewhere.

I am involved in a new company in west Cork that will take advantage of excess farm manure in this way. The Celtic Worm Company has resulted from an investment of forty farmers (and growing) and from business startup funding from the West Cork Development Partnership. We aim to provide the farmers with the skills and equipment necessary to make a top-class, fully mature compost that will comply with the compost quality assurance scheme. We will then buy the excess compost from the farmer for the purpose of either vermicomposting it or leaving it as is for use in our Celtic Gold range of plant foods and compost blends.

Having a company that, in essence, belongs to the community has brought a great buzz to the area. And having so many people involved results in ample innovative ideas, expertise, resources and enthusiasm. I hope this can become a pilot for creating more such organisations.

In Germany and other parts of Europe, this type of collaboration is commonplace. For instance, farm partnership schemes allow farmers to get together and run their farms jointly. With this type of scheme, they can bulk buy and reduce their costs. They can also share machinery and land, and pool their skills, reducing costs even further. As well as a financial benefit, there are social benefits also. Having two or more people working together on a farm allows for time off for things like holidays (almost a foreign word to farmers today) and illness. And we all know that more time spent at home allows for a better work–life balance. There is a lot of work still to be done in Ireland to make these types of partnerships more attractive, but the benefits are clear for all to see.

With the Celtic Worm Company, I hope to achieve at least some of the same results. This system can be rolled out across the country and set up in other rural communities – and not just for composting but also anaerobic digestion, fuel creation and food production industries. When communities get together, more can be achieved, costs are spread out, purchasing power is greater, skills are broader and there are more contacts. It is both an environmentally and socially sustainable way of working – and surely it is a step in the right direction in the wake of the Celtic Tiger.

Notes

1 Davies, A. *et al.* (2005), 'Mind the Gap! Householder Attitudes and Actions Towards Waste in Ireland', *Irish Geography*, 158.

Éabha Ní Laoghaire Nic Ghiolla Phádraig

A Vision of Ireland (Translation of Chapter 2)

When Éabha was younger, she wanted to be a professional saint. On realising that wasn't possible, she then aimed to be the first Irishwoman in space. In the end, though, it was her love of languages that won out: she chose to study French and Irish in the University of Limerick, from which she graduated with first class honours in 2012. Now completing an MSC in Innovation, Commercialisation and Entrepreneurship at University College Cork, she claims she 'accidentally fell in love with Irish' thanks to her uncle and also to a great supervisor in college. She was highly commended in the Irish Language category of The Undergraduate Awards in 2012.

A few days before I wrote this essay, my grandmother Hancy Fitzpatrick sadly passed away. She died just short of her ninety-seventh birthday, a remarkable feat given the great hardship that punctuated her life. Perhaps the foremost example of this hardship was the closure of the family business during the Economic War of the 1930s. For my grandmother, the closure was compounded by the death of her husband, which meant that she had to raise nine children alone.

In order to provide for these children, my grandmother put herself through the demanding routine of working night shifts in a hospital before coming home in the morning to prepare breakfast. This would be followed by four hours' sleep, after which dinner was made and she would then return to the hospital for a night of more work. But my grandmother could always count on the support of her neighbours as she admirably and tirelessly devoted her life to her family. Indeed, despite her unenviable circumstances, she would regularly welcome people into her home for social occasions. She even found time to set up the Cork branch of the National Association of Widows in Ireland with Maureen Black, and co-founded the Community Centre in Blackpool – an initiative for the local community that granted access to medical facilities.

Given that we are currently in the midst of our own economic crisis, I often find myself thinking what my ancestors, including my grandmother, would do to cope with and overcome the situation we currently find ourselves in. It is my belief that answers to these compelling questions reside in Irish literature. Is there wisdom and sagacity to be found in Irish literature that can guide and inspire us in overcoming the country's demise? In this essay I will make reference to some Irish works that I feel teach us important lessons that could prove influential in creating a positive vision for our country.

THE POWER OF IRISH LITERATURE

I often remember swallowing back tears in the middle of a lecture, experiencing such strong emotions in sympathy with the characters portrayed in famous works of Irish literature, so moved was I by their pain, terror, courage, strength, defiance and disruptiveness. Equally, other works had the capacity to reduce me to tears of laughter. A variety of emotions have been stirred in me by the poetry and stories based on events that have occurred in this country. This has had a profound effect on the way I perceive Irish history, as well as Ireland's current situation. I have a deep enjoyment and appreciation of English and French literature but my affinity for Irish literature will always be unrivalled, such is the connection between the literature, our history and our ancestors.

Identity is a prominent theme in much of Irish literature and, in experiencing these works, I can understand more clearly the experiences of my ancestors and the essence of the Irish mentality. One of the traits I love about Irish literature generally is that it focuses on the ordinary, everyday people, whose exemplary attitude and mindset can provide valuable lessons for us today.

THE IMPORTANCE OF COMMUNITY

Tomás Ó Criomhthain's autobiography entitled *An t-Oileánach* (*The Islandman*) describes how life on the Blasket Islands at the end of the nineteenth century and the beginning of the twentieth century depended so much on community and resourcefulness. The phrase '*Níl ann ach an fhírinne; níor ghá dom aon cheapadóireacht*' ('It is only the truth; there was no need to compose')[1] points to the authenticity of Ó Criomhthain's description of life on the island. It is important to understand that Tomás always speaks on behalf of the community, as indicated by the phrase *'Thugas iarracht ar mheon na ndaoine do bhí im thimpeall do chur síos chuin go mbeadh a dtuairisc'* ('I tried to give a description of the mindset of the people who were around me').[2] As a result, throughout the book Ó Criomhthain throws light on the hardship of life on the island: *'Oíche mhór fhada fhuar mar seo ag comharc na mara, go lánmhinic ar bheagán fáltais'* ('A great long cold night foraging food and often with little to show for it').[3]

One such remarkable incident that highlights the importance of community and resourcefulness is *'An Long Gail is an t-Arm'* ('The Steam Ship and the Army'), where soldiers come to the island in order to collect rent from the inhabitants. The inhabitants' greatest fear is that these soldiers will steal their animals and destroy their homes. Immediately, the community congregate to prevent the imminent danger and preserve their way of life. *'Fuaireas post ós na mná do bhí bailithe ann. Ag bailiú chloch do cuireadh me agus gach nduine eile do bhí ann'* ('I was given a job by the women who had congregated of collecting stones like everyone else').[4] The men left while the women and children collected stones and headed for the harbour: *'Ná raibh na mná ag fágáil na háite'* ('The women were not leaving').[5] The women and children were unwilling to leave their beloved island undefended, highlighting their renowned strength and resourcefulness.

Approaching the congregation on the beach were three boats with trained, armed soldiers who could have caused havoc on the island. Despite the looming danger, however, *'ní raibh eagla ar na mnáibh rompu'* ('the women were not afraid').[6] In fact, the women and children started hurling stones at the soldiers as they descended the ship, badly injuring them. There was even one woman so determined to fight off the soldiers that she was willing to sacrifice her own child to protect the island; in a moment of madness she almost threw her baby: *'go gcaithead an leanbh leo!'* ('that she would throw her child!').[7] Even without sufficient resources or ammunition, these islanders gathered together to ensure the community's survival and, as a result, *'Do ghluaisíodar leo abhaile insan imeacht go dtánadar, gan bó, gan capall, gan caora'* ('They [the soldiers] left without any cows, horses or sheep').[8]

Co-ordinating successfully as a unified community and making use of all the resources available, the islanders won and defeated the soldiers. Perhaps there is a lesson in this that we could learn. Certainly, the people of the Blasket Islands have taught me that greatness and determination lies within the Irish people: as a united community, we are unstoppable even against the greatest of foes.

I wholeheartedly believe that it is essential to support each other and to function as a unified community. Referring back to my grandmother's story, it was the community that helped and supported her. Our ancestors had the capability to come together as one in order to overcome incredibly difficult obstacles. As direct descendants of these formidable people, we too can congregate to surpass our current economic hurdle. It is written in history and evident in our literature that Irish people are robust, tenacious and tough. We have proven this time and time again.

DISRUPTIVE BEHAVIOUR: BOLDNESS AND DEVIANCE

My grandmother always insisted on standing up for people's rights. It is essential to express your voice and to contest social norms and values that wrongfully restrain people. This is a recurring theme in the poetry of Nuala Ní Dhomhnaill.

In her poem '*Táimid Damanta, a Dheirféaracha*' ('We are Damned, My Sisters'), published in 1988, Ní Dhomhnaill makes reference to Eve. She does this in several other poems, including '*Manach*' (1998) and '*Cnámh*' (1988). Eve symbolically represents deviance and disruption. She is the root cause of original sin: it was Eve who seduced Adam into eating the apple in the Garden of Eden. Readers get the distinct impression that Ní Dhomhnaill supports and defends Eve's implied boldness and daring:

Chaitheamar oícheanta ar bhántaibh Párthais
ag ithe úll is róiseanna
laistiar dár gcluasa ag rá amhrán
timpeall tinte cnámh na ngadaithe
ag ól agus ag rangás le mairnéalaigh agus robálaithe.[9]

We spent nights in Eden's fields
eating apples, gooseberries; roses
behind our ears, singing songs
around the gipsy bon-fires
drinking and romping with sailors and robbers.

The women in '*Táimid Damanta, a Dheirféaracha*' are wholeheartedly indifferent to social rules and norms. This sense of indifference is very powerful: daring behaviour administers these women total authority over their lives. Similar themes are to be found in other works by Ní Dhomhnaill. In *'Labhrann Medb'* and *'Agallamh Na Mór-Riona Le Cú Chulainn'* Ní Dhomhnaill highlights daring and unorthodox behaviour and, subsequently, the empowerment of women.

Ní Dhomhnaill gives prominence to the expression of one's voice. The characters in her poems have a great capacity to express themselves – indeed, they cannot be silenced. As Ní Dhomhnaill herself exclaims: 'It is awful to be invisible.'

These characters have had a remarkable impact on me and their profound message resonates deeply. It is evident from Ní Dhomhnaill's poetry that disruptive behaviour is essential; we must be bold like these characters. In the middle of an economic crisis, when society has been turned upside down and inside out, is it not imperative to be courageous and daring? Are we not obliged to speak out against social injustices that damage our growth and our development as a distinguished nation?

QUESTIONING AND CHANGING SOCIAL NORMS: BREAKING FREE FROM TRADITIONAL VALUES

Ní Dhomhnaill certainly portrays the importance of being fearless. Equally, she illustrates the need to give people a voice. As a result, we are encouraged to question the conditions of society. In *'Táimid Damanta, a Dheirféaracha'*, Ní Dhomhnaill connects the immoral character Izeibil with the women who do not abide by the role and status assigned to them within the patriarchal system.

Ní Dhomhnaill's poetry requires a consideration of the philosophy of existentialism. Earnshaw said: 'Existence precedes essence … man being essentially "nothing" but what he makes of himself.'[10] I believe there is an element of existentialism in '*Táimid Damanta, a Dheirféaracha*'.

The existential concepts of *être-pour-soi* (being-for-itself) and *transcendence* are important concepts when making an analysis of the characters in '*Táimid Damanta, a Dheirféaracha*'. A person who practises *être-pour-soi* does not abide by social rules nor adhere to social norms, especially if these rules and norms prevent or restrict freedom. Instead, this person questions the state of society, social roles and the power of authorities. According to Jean-Paul Sartre, it is our duty to accept individual responsibility for the state of society and condition of life, and this is understood through a consciousness of existence. The aim of *être-pour-soi* is for humankind to achieve freedom; in doing this, we must accept total responsibility for the state and condition of society as it currently is. This concept teaches us that it is insufficient to put blame onto society, the state, the church or any other authoritative institution. It declares that we, as individuals, have the power within us to change the world.

In existentialist terms, *transcendence* dictates that a person is not restricted by their social class, social roles, social norms or values. According to Sartre, the status or the role that a person has acquired in society is immaterial; they still have the potential to break free, since they are neither restricted nor limited in any way by this role or status.

With reference to both *être-pour-soi* and *transcendence* in '*Táimid Damanta, a Dheirféaracha*', the women of this poem do not adhere to the patriarchal social structure. They are indifferent to the role that is imposed upon them by society; they are unwilling to accept the lesser status of women or their condition of life: *'Ná bheith fanta/Istigh ag baile ag déanamh tae láidir d'fhearaibh'* ('than to be indoors making strong tea for the men').[11] These women are not deterred by class, social roles or values. They break free from any limitation, totally unconcerned with their status. They are willing to enjoy the pleasures of life: *'Is rince aonaoir a dhéanamh ar an ngaineamh fliuch'* ('dancing singly on the wet sand').[12] This could almost be seen as an affront to societal norms.

These women share their voices of strength, resilience and defiance; they are indomitable, insurmountable and courageous. They are determined to follow the uneasy path to freedom, alleviation and ecstasy and they, in effect, beseech us to question the state of society and to shed light on the norms and traditions that no longer serve us. The message of this poem reverberates in every sense of my being. I feel that, like the characters in this poem who made a conscious effort to make changes in society, we too have the potential to highlight key issues. We have the opportunity to reverse any unfair or detrimental direction society may have pursued. We have the possibility to find solutions that will create a magnificent and harmonious Irish society.

ENJOY! THE IMPORTANCE OF ATTITUDE

The autobiography *Fiche Bliain ag Fás* — (*Twenty Years A-Growing*) by Muiris Ó Súilleabháin, published in 1933, begins by outlining the hardship of his youth: *'Ná rabhas ach leathbhliain d'aois nuair a fuair mo mháthair bás... ní raibh éinne chun aire a thabhairt domsa'* ('I was only six months old when my mother died...there was nobody there to take care of me').[13] Despite this horrific ordeal, his writing demonstrates the bliss he managed to experience throughout everyday life: *'Ní fada go rabhas i dtaibhreamh, go rabhas féin agus Micil Dé ag siúl trí pháirc leathan i mBaile an Mhuilinn ag baint bláthanna deasa'* ('It wasn't long till I was dreaming, and Micil Dé and I were walking through a wide field in Miltown picking lovely flowers').[14] Quotes like this show that Ó Súilleabháin had a very special attitude to the world around him – a zest and joy for life shines through his writing. In addition to this, there is an evident link between his positive mindset and this creativity and innovation. For example: *'"Ó," deireadh sí, "tabharfaidh mé isteach in áit dheas inniu thú." – "Agus an bhfuil milseáin ann?" arsa mise.'* ('"Oh," she said, "I will take you to a nice place today." – "Are there any sweets there?" I said.').[15] Clearly, Ó Súilleabháin grasps every opportunity that presents itself to him; even at a very young age, he thinks outside the box in a humorous and creative way.

As with the people of the Blaskets, Ó Súilleabháin enjoyed everyday life. In the chapter *'Rásieanna Fionntrá'* ('Ventry Races'), he gives another example of having an open mindset and embracing opportunity. He and his friend Tomás go to the races and, in a very funny incident, end up getting drunk, throwing up and innocently buying sweets on the way home. *'Cá mbeimís ach ar aghaidh thí óil amach i gCeann Trá?'* ('Where would we be except in front of a pub out in Ventry?').[16] In the chapter *'Oíche Shamhna'* ('Halloween'), Ó Suilleabháin gives a detailed description of the entertainment and games they created; we are told that *'árdoíche le bheith againn'* ('a great night was to be had').[17] These chapters show a link between creativity, innovation and fun. Ó Suilleabháin and his friends wholeheartedly took the opportunities that were presented to them to try something new and inevitably they really enjoyed themselves! This is a valuable lesson for all of us.

As Ó Súilleabháin illustrates, such events epitomise the joys of living; he never took life too seriously, even though he had many good reasons to do so. The characters enjoyed themselves. It is part of the Irish condition: we can be whimsical and witty. Ó Súilleabháin emphasises the importance of a positive outlook and an attitude of wonder in life. It reminds me of my grandmother's story and the parties she would host in her little house. Despite life's hardships and turbulences, it is essential to cherish and conserve a positive and extraordinary mindset. In short – enjoy life!

This capacity for joy is a trait that my grandmother embodied and practised until the day she died. This was her vision for the country: communities would congregate and work together as one; society would emphasise the importance of resourcefulness; people would make full use of the means available to them, without complaining about what they lacked.

This same zest for life can be seen all throughout Irish literature. We see it in the stories of the people of the Blasket Islands. In Nuala Ní Dhomhnaill's poetry, we see the importance of being bold – sometimes disruptive.

I can imagine my own grandmother saying: 'Do not accept life or the current condition of the country if you do not agree with it.' She often used to challenge me: 'Don't just *complain* if people are suffering. You must express your voice – speak out!' And she was right. Just like the society in '*Táimid Damanta, a Dheirféaracha*', Ireland has social issues that need to be addressed. If people are suffering, then there is a need for change.

Of course, my grandmother would also agree with Ó Súilleabháin's mindset: life is not inherently sombre or hopeless. She once said some very powerful words to me: 'Enjoy life, then creativity and innovation will follow.'

My grandmother's many words of wisdom continue to echo in my heart. She was right to recognise Ireland's potential. We have an innate aptitude for creating long-lasting, tangible solutions that will improve the state of the country and create a better future for us all and for those who are yet to come.

Quotation and title translations are author's own unless otherwise stated.

Notes

1 Ó Criomhthain, T. (2002), *An t-Oileánach.* Baile Átha Cliath: Cló Talbóid, p.325.
2 Ibid. 327.
3 Ibid.
4 Ibid. 50.
5 Ibid. 51.
6 Ibid.
7 Ibid. 52.
8 Ibid. 53.
9 Ní Dhomhnaill, N. (1988), *Selected Poems: Rogha Dánta,* tr. Michael Hartnett. Baile Átha Cliath: New Island, p.14.
10 Earnshaw, S. (2006), *Existentialism: A Guide for the Perplexed.* London: Continuum International Publishing Group, p.74.
11 Ní Dhomhnaill, N. (1988), *Selected Poems: Rogha Dánta,* tr. Michael Hartnett. Baile Átha Cliath: New Island, p.15.
12 Ibid.
13 Ó Súilleabháin, M. (2011), *Fiche Bliain ag Fás.* An Daingean: An Sagart, p.11.

14 Ibid. 19.
15 Ibid. 11.
16 Ibid. 74.
17 Ibid. 54.